Technological Change in Japan's Beef Industry

Westview Special Studies

The concept of Westview Special Studies is a response to the continuing crisis in academic and informational publishing. Library budgets for books have been severely curtailed. Ever larger portions of general library budgets are being diverted from the purchase of books and used for data banks, computers, micromedia, and other methods of information retrieval. Interlibrary loan structures further reduce the edition sizes required to satisfy the needs of the scholarly community. Economic pressures on the university presses and the few private scholarly publishing companies have severely limited the capacity of the industry to properly serve the academic and research communities. As a result, many manuscripts dealing with important subjects, often representing the highest level of scholarship, are no longer economically viable publishing projects--or, if accepted for publication, are typically subject to lead times ranging from one to three years.

Westview Special Studies are our practical solution to the problem. We accept a manuscript in camera-ready form, typed according to our specifications, and move it immediately into the production process. As always, the selection criteria include the importance of the subject, the work's contribution to scholarship, and its insight, originality of thought, and excellence of exposition. The responsibility for editing and proofreading lies with the author or sponsoring institution. We prepare chapter headings and display pages, file for copyright, and obtain Library of Congress Cataloging in Publication Data. A detailed manual contains simple instructions for preparing the final typescript, and our editorial staff is always available to answer questions.

The result is a book printed on acid-free paper and bound in sturdy library-quality soft covers. We manufacture these books ourselves using equipment that does not require a lengthy make-ready process and that allows us to publish first editions of 300 to 1000 copies and to reprint even smaller quantities as needed. Thus, we can produce Special Studies quickly and can keep even very specialized books in print as long as there is a demand for them.

About the Book and Authors

This book evaluates Japan's ability to meet domestic beef needs for the next quarter century and considers prospects for the Japanese beef industry. It also provides an in-depth review of production practices from the viewpoint of technological and institutional change. Emphasis is given to the impact that new technologies such as embryo transfer and growth hormones will have on production costs and to the social consequences of adopting new technology. The authors conclude by comparing the recent experiences of Japan to changes taking place in the United States and the nations of the European Economic Community.

James R. Simpson is professor and livestock marketing economist in the Food and Resource Economics Department at the University of Florida. Tadashi Yoshida is associate professor in the Department of Agricultural and Forestry Economics at Kyoto University, where Ryohei Kada is assistant professor. Akira Miyazaki is associate professor in the Department of Animal Science at Kyoto University.

Technological Change in Japan's Beef Industry

James R. Simpson, Tadashi Yoshida, Akira Miyazaki, and Ryohei Kada

Routledge
Taylor & Francis Group
LONDON AND NEW YORK

First published 1985 by Westview Press, Inc.

Published 2019 by Routledge
52 Vanderbilt Avenue, New York, NY 10017
2 Park Square, Milton Park, Abingdon, Oxon OX14 4RN

Routledge is an imprint of the Taylor & Francis Group, an informa business

Library of Congress Cataloging in Publication Data
Main entry under title:
Technological change in Japan's beef industry.
(Westview special studies in agricultural science and policy)
1. Beef industry--Japan--Technological innovations.
I. Simpson, James R.
HD9433.J32T43 1985 338.1'76213'0952 84-15260

ISBN 13: 978-0-367-28958-4 (hbk)

Contents

Photographs

Photographs follow page 126.

Plate 1. Wagyu cows have traditionally been raised on very small operations and spend virtually all their lives confined in a barn.

Plate 2. Modern Wagyu cow/calf operations have 10-20 cows and heifers. They are permitted access to loafing areas even in the upper half of Japan which is covered with snow 4-5 months per year.

Plate 3. Wagyu calves are carefully groomed prior to entering the auction ring. Virtually all Wagyu calves are sold through auctions.

Plate 4. Wagyu calves are sold individually at auction. Potential buyers have access to extensive pedigree information even though most calves are sold for feeders.

Plate 5. Modern feedlots for Wagyu have sprung up all over Japan. There is no stocker period on pasture so Wagyu are maintained in a confinement feedlot from weaning to slaughter.

Plate 6. Crossbreeding to produce feeder calves or for development of a composite beef breed offers great potential to expand productivity. But, adoption of this production technique will likely be slow in coming.

Plate 7. Dairy cattle are maintained in a confinement feedlot from about four months of age to slaughter. Bedding, changed every 7-10 days, is used in most operations.

Plate 8. There are a wide variety of feedlot facilities. Producers, especially of dairy steers, are continually experimenting to reduce cost and increase gain.

Preface

The Japanese Cattle industry has been undergoing major changes for the past three decades. During the 1950's and 1960's mechanized power rapidly replaced cattle and horses as draft power. By the middle 1960's--just 20 years ago--that process was complete. Then, during the middle to late 1960's, and on into the early 1970's attention was placed on improvement of native beef breeds, primarily Wagyu. At the same time the dairy industry developed to the extent that by the middle 1970's it was the principal source of domestically produced beef.

The process of beef industry structure change accelerated in the 1970's as medium scale feedlots came into being, regional packing plants were established, and the beef marketing system matured. Economic forces, both within and external to the industry, resulted in a major shift by farmers out of cattle raising and associated rapid increases in average farm size. In the mid-1960's there were 1.4 million farms with cattle. By 1975 there were only half as many, and by 1984 only about 315,000 farms with cattle--one-fifth the number 20 years previous. Most important, the decline in farms raising cattle is continuing at the same rate in the 1980's as in the 1970's

The cattle industry, and beef breed cow/calf operations in particular, have played an extremely important role in Japan, mainly by providing an income source for rural people with few employment alternatives. There are many reasons why analysis of the cattle industry is important, but perhaps most compelling is the need to evaluate the role which will be played by small scale raisers in the future. It was recognition of structural changes which led Dr.'s Yoshida, Miyazaki and Kada to form a Japan cattle industry working group headquartered at Kyoto University.

This book, <u>Technological Change in Japan's Beef Industry</u>, is one product of this effort which has become international in scope.

A major objective of this book is to test our hypothesis that beef production by Japan's cattle industry could become competitive with imported beef. This hypothesis is contrary to general belief both within and outside Japan that, because of land constraints and structural characteristics, import liberalization (or at least a major step toward it) would result in demise of Japan's beef industry. As part of the analysis we argue that beef must be viewed as a highly differentiated commodity on both the supply and demand sides--and that the amount of differentiation will continue. In effect, we vigorously counter the notion held by virtually all foreigners that beef is high priced in Japan. The fact is that in certain restaurants and in some meat markets most of the beef sold is not much higher priced than equivalent quantity in countries exporting to Japan when transportation, duty and other marketing charges are subtracted.

Another objective of this book is to highlight the dramatic changes which have taken place in Japan's cattle industry and to emphasize how and why changes will continue to take place. We hypothesize--again in opposition to received belief--that Japan's beef industry has more potential for development than is generally recognized. One reason for this conclusion is the paradox that because production is on a very intensive scale there is both incentive and opportunity to employ many technologies not now being used. In this book we tie technology evaluation with adoption potential.

The Japanese experience in transformation of its livestock industry in a very short period has great relevance for many countries, especially those in Asia-primarily ones that are land and labor intensive. Thus, one purpose of this book is to analyze policies linked to the modernization process. The cornerstone of analysis is review of technologies which are either now available, or will be available in the next two decades. Various scenarios are developed to aid in estimating the potential impact. These scenarios should not be construed as predictions, but rather as a means to examine the potential impact on productivity and cost. The scenarios are specific to Japan, but have relevance to evaluate potential impact of technological change on the cattle industry of any country.

Many people and institutions have helped to make this book possible. Thanks go to our employers, Kyoto University and the University of Florida for recognition of benefits from international collaboration. We are indebted to Professors Taiji Kikuchi, Ryoji Kawashima and Taira Yori of Kyoto University as well as Seiji Mori, Satoshi Ohara, James Echle, Robert Railsback, Yoshihiro Yamashita, and the Ito Foundation for their support. Yoshiaki Iiguni and Tsutomu Tanaka, graduate students at Kyoto University and Hiroshi Tanaka, former graduate student at the University of Florida provided invaluable support in data preparation. Our profound thanks go to Marianne Mangone and Terri Parsons for typing manuscripts and Toni Glover for final draft preparation.

A curious event happened in Japan during the Meiji Era which helps set the tone to understand technological change in Japan's beef industry over the next quarter century. It seems that during 1876-1877 an American, Dr. William S. Clark, President of Massachusetts Agricultural College spent eight-and-a-half months in Hokkaido, now a major livestock production area of Japan, as President of Japan's first agricultural college. Although reference to him can scarcely be found in the United States, Dr. Clark's name and story are familiar to even the youngest Japanese student. This was partly because he was so successful in his pioneering work--but more important has been his parting words. Dr. Clark's last day was spent in the country with his twenty-four students. After lunch and a walk around the area he mounted his horse, turned to them and shouted "Boys, be ambitious!" He rode off, never looking back. Our parting words nearly 100 years later--at a time when international trade conflicts and societal change demand deep understanding of Japan's beef industry--might well be "Boys, be imaginative!"

James R. Simpson
Tadashi Yoshida
Akira Miyazaki
Ryohei Kada

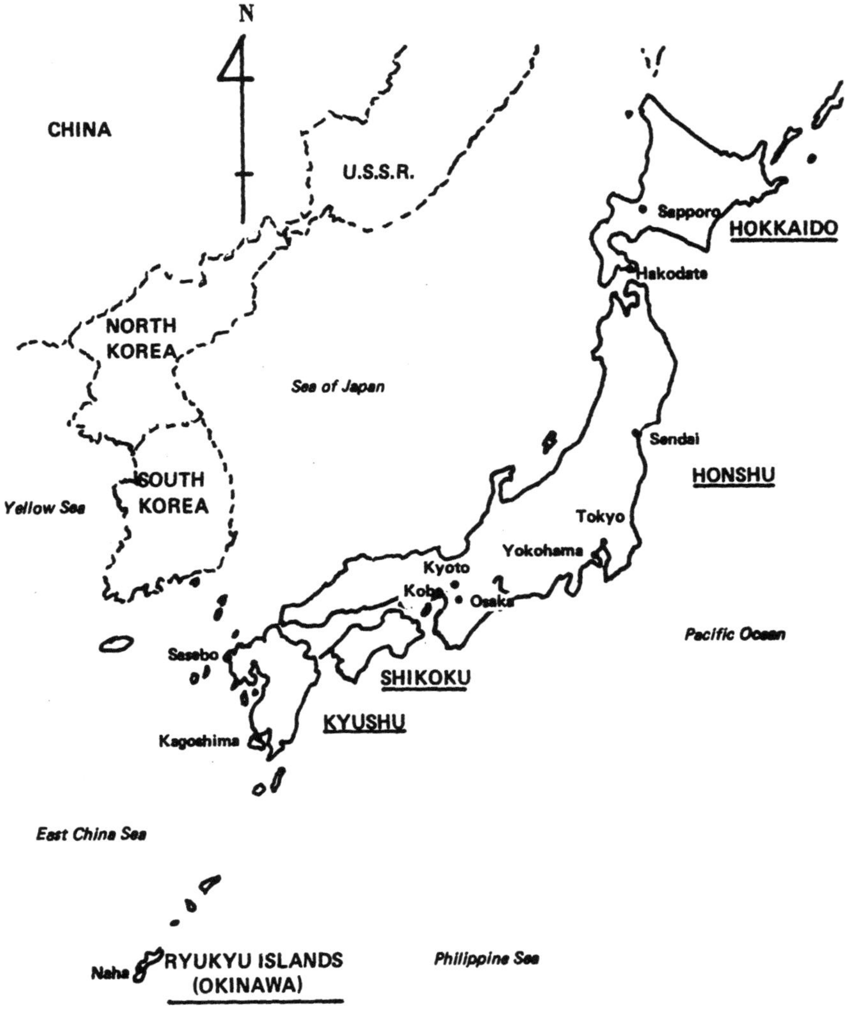
N
CHINA
U.S.S.R.
NORTH KOREA
SOUTH KOREA
Yellow Sea
Sea of Japan
Sapporo
HOKKAIDO
Hakodate
Sendai
HONSHU
Tokyo
Yokohama
Kyoto
Kobe
Osaka
Pacific Ocean
Sasebo
SHIKOKU
KYUSHU
Kagoshima
East China Sea
Naha
RYUKYU ISLANDS (OKINAWA)
Philippine Sea

JAPAN

Part I

Industry Structure and Dietary Patterns: Introduction to Complex Interrelationships

1 Introduction to Japan's Agricultural Economy

Japan's economic development has been described as a modern day "miracle" by some writers. Part of any nation's development process is a change in tastes and preferences for food products as well as shifts from the so called "inferior" goods to those considered "superior." Throughout the world one of these shifts has traditionally involved the substitution of meat for starches. That additional meat can either be produced domestically by reallocation of resources, or imported.

The major purpose of this book is to provide the relevant facts, figures and analysis needed to understand the impact that technological change in cattle production will have on Japan's beef business. We begin by providing a sketch of agriculture in Japan. Then, in the next chapter, color is added to the picture by focusing on the changes which have and are taking place in dietary patterns and beef consumption. Part II deals with the current production situation. The technological changes which are taking place and will have an impact are the subject of Part III. The overall objective is to analyze changes which are taking place--and will take place--as a basis to understand beef production in Japan.

POPULATION PRESSURE ON LAND USE

Total cultivated land area in Japan has decreased about 9 percent in the past two decades mainly as a result of urbanization, industrialization and expansion of the highway network. Today, only 13.1 percent of all land area is classified as being arable, compared with 15 percent in 1965. Two thirds of all land is in forests, while about one fifth is in other uses (Appendix 1.1). Approximately 55 percent of all available arable land is in rice paddy fields and 33 percent in other crops. In other words, there is relatively little land for extensive cattle production.

The data on Japanese land use is placed in better perspective by a comparison with other countries. For example, Japan is similar to Korea by having two thirds of total area in forest and woodland. However, despite only 13.1 percent of Japan's land area being arable, that percentage is greater than the world's average of 11.2, Canada's 5.0, China's 10.8 or Australia's 5.7 percent (Appendix 1.1).

The information about land use is interesting, but more illuminating is the impact from very heavy population pressure. There are now about 4.7 billion people in the world of which about 45 percent are classified as agricultural. Japan has 119 million people but is only the size of California. This small country is somewhat unusual among the developed countries in having 9.0 percent of total population classified as agricultural. Comparable figures are 1.9 percent in the United States, 5.2 percent in Australia, and 3.3 percent in West Germany (Appendix 1.2).

Japan's population of 119 million people is roughly 51 percent that of the United States. The population growth rate of Japan has slowed down from 1.33 percent annually during the 1970-75 period to an estimated 0.57 percent from 1980 to 1985. The rate is expected to drop to 0.40 by the end of the century and reach equilbrium about 2010. Despite these encouraging trends, data on total population pales in comparison with density calculations. Japan, with 322 people per square kilometer (km^2) has, by far, one of the highest densities of any medium or large country in the world (there are a few countries, territories or protectorates such as Puerto Rico or Hong Kong with much higher densities). In contrast, West Germany has 251 people per km^2 while Canada has 3, the United States 26, and China 111 (Appendix 1.2).

The pressure on land resources is dramatized even further when population density is related to arable land and permanent crops. Using this definition, Japan has 2,458 people per km^2, West Germany 821, Canada 54, the United States 124 and China, a country that is generally thought of as being very densely populated, "just" 1,024. The world average is 318 people per km^2 of arable and permanent crop land.

FARM STRUCTURE

The proportion of Japan's population considered to be in agriculture has declined continuously over the past few decades, from 30.6 percent in 1960, to 9.0 percent in 1983 (Appendix 1.3, FAO data) and that rate is continuing to fall. The proportion of farms to total households has also declined continuously, from 23.5 percent in 1965, to 11.8 percent in 1984 (Table 1.1). The rate of decline has been 3.6 percent annually for the whole period. It has averaged 2.4 percent annually in the 1980's. The population living in farm households exceeds population considered as agricultural since many members of farm families live in rural households, but work in nonagricultural activities. The salient point is that Japan is continuing to urbanize and the number of farm households will probably continue to decline in both absolute and relative terms.

Table 1.1. Total households and farm households in Japan, 1965-1984

Year	Total households in Japan	Farm households	Proportion farm to total households
	- - - - 1,000	- - - -	- Percent -
1965	24,082	5,665	23.5
1970	28,093	5,402	19.2
1975	32,141	4,953	15.4
1980	36,015	4,661	13.0
1981	36,347	4,614	12.7
1982	36,859	4,567	12.4
1983	37,426	4,522	12.1
1984	37,935	4,473	11.8

Source: MAFF. <u>Abstract of Statistics on Agriculture, Forestry, and Fisheries, 1984</u>. Tokyo, March, 1985.

Japan is divided, for agricultural purposes, into two parts. One, which might be called the older or "original," comprises the islands of Honshu, Shikoku and Kyushu. Within those areas the major regions were Kinki, Tokai, Kanto and Sanyo. Since World War II, and in the last 20 years in particular, the Kyushu and Tohoku regions have become major agricultural producers (Figure 1.1). This has mainly been due to change from industrialization of the country.

The other part of Japan (apart from Okinawa) is the northernmost island of Hokkaido, an area that has rapidly developed since World War II and is characterized by much larger land holdings than the "original" area. Hokkaido, while at a latitude similar to Wisconsin or Massachusetts in the United States, has a much milder climate because of being surrounded by water. The result is a sufficiently long growing season to be a major agricultural area.

Total farm households in the prefectures excluding Hokkaido have fallen from 5.2 million in 1970 to 4.4 million in 1984, an average decline of 1.3 percent annually (Appendix 1.4). All of that decline has taken place in farms with 2 hectares or less, as farms with more than 2 hectares have grown slightly. On a proportion basis, farms with more than 2 hectares have grown from 5.8 percent of total farms in 1970 to 8.4 percent in 1984 (Appendix 1.5). The structural changes have been even more dramatic in Hokkaido, where total number of farm households dropped from 166 thousand in 1970 to 113 thousand in 1984. Farms with 20 hectares or more increased from 3 percent of the total in 1970 to 14.2 percent in 1984. Overall, it appears the trend in both parts of Japan toward fewer and larger farms is continuing although at a somewhat slower rate than in the 1970's.

A substantial change has also taken place in employment for, while 34.3 percent of all farm households were considered "full-time" in 1960, by 1975 that classification had dropped to 12.4 percent of the total. Since then the full-time proportion has increased steadily but slowly, reaching 13.5 percent in 1984 (Table 1.2). A closer look at the part-time segment shows that about 82 percent of them have their members mainly engaged in other jobs. Perhaps the most dramatic statistic of all is that, of the 4.5 million farm households in Japan (which comprise about 20.5 million people or about 17 percent of total population) only about 5.6 million people, or only 4.7 percent of Japan's entire population, are engaged exclusively in farming (Appendix 1.6 and Table 1.2).

Further examination of data on occupation of family members reveals that while in 1970 an average of 1.63 farm family members per household were engaged in farming, by 1983 the average had fallen to 1.15 persons

Table 1.2. Full-time versus part-time farm households, Japan, 1960-1984

				Principal occupation of part-time households	
Year[a]	Total farm households	Full-time	Part-time	Mainly engaged in farming	Mainly engaged in other jobs
	1,000				
1960	6,057	2,078	3,979	2,036	1,942
1970	5,402	845	4,557	1,814	2,743
1975	4,953	616	4,337	1,259	3,078
1980	4,661	623	4,038	1,002	3,036
1981	4,614	580	4,035	829	3,205
1982	4,567	599	3,968	774	3,194
1983	4,522	596	3,926	731	3,195
1984	4,473	605	3,868	689	3,179
	Percent				
1960	100	34.3	65.7	51.2	48.8
1970	100	15.6	84.4	39.8	60.2
1975	100	12.4	87.6	29.0	71.0
1980	100	13.4	86.6	24.8	75.2
1981	100	12.6	87.4	20.6	79.4
1982	100	13.1	86.9	19.5	80.5
1983	100	13.2	86.8	18.6	81.4
1984	100	13.5	86.5	17.8	82.2

Source: MAFF. Abstract of Statistics on Agriculture, Forestry, and Fisheries, 1984. Tokyo, March, 1985.

[a]Includes Okinawa beginning in 1970.

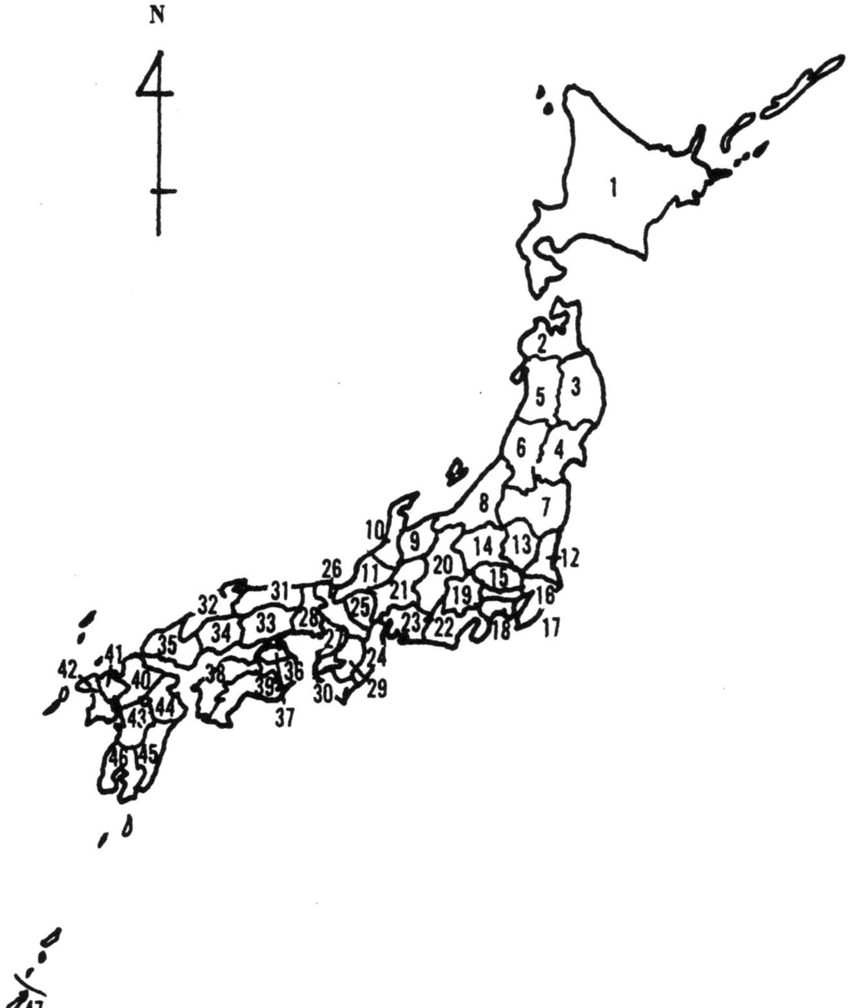

Figure 1.1 Prefectures in Japan

Map Number	Agricultural Region and Prefecture	Map Number	Agricultural Region and Prefecture
	Hokkaido		Kinki
1	Hokkaido	25	Shiga
		26	Kyoto
	Tohuku	27	Osaka
2	Aomori	28	Hyogo
3	Iwate	29	Nara
4	Miyagi	30	Wayayama
5	Akita		
6	Yamagata		Chugoku
7	Fukushima		
			Sanin
	Hokuriku	31	Tottori
8	Niigata	32	Shimane
9	Toyama		
10	Ishikawa		Sanyo
11	Fukui	33	Okayama
		34	Hiroshima
	Kanto-Tosan	35	Yamaguchi
	Kita-kanto		Shikoku
12	Ibaraki	36	Tokushima
13	Tochigi	37	Kagawa
14	Gumma	38	Ehime
		39	Kochi
	Minami-kanto		
15	Saitama		Kyushu
16	Chiba		
17	Tokyo		Kita-kyushu
18	Kanagawa	40	Fukuoko
		41	Saga
	Tosan	42	Nagasaki
19	Yamanashi	43	Kumamoto
20	Nagano	44	Oita
	Tokai		Minami-kyushu
21	Gifu	45	Miyazaki
22	Shizuoka	46	Kagoshima
23	Aichi		
24	Mie		Okinawa
		47	Okinawa

(Appendix 1.6). Furthermore, the number of those only engaged in farming their own operations (called self-farming) dropped from 1.44 members to 1.03. Total labor time per family has also declined dramatically from 2,661 hours per year in 1970 to 1,932 in 1983. Alternatively, there was a drop from about 51 hours to 37.3 hours per week on a 52 week per year basis. These are all indications of the economies from larger size operations, the continued introduction of more efficient machinery during this period, and a shift in crop and animal mix. As might be expected, because the farms are very small, little use is made of hired labor with it accounting for only about two percent of total hours worked. Most of this is in the harvest period or on commercial dairy, poultry and swine enterprises.

Considerable change has also taken place over the past 10 years in types of agricultural commodities produced. Total rice output in the mid-1970's has now stablized at about 70 percent of that peak (Appendix 1.7). Wheat and barley production, although never very large, are now more than double the 1975 level. Vegetable, fruit and nut production have increased faster than population. Sericulture, once a major cottage-type industry, has declined considerably, and output is now only half what it was in 1970.

INTERNATIONAL TRADE

The total value of Japan's exports increased from 8.5 billion U.S. dollars in 1965 to $168.3 billion in 1984, although it had declined to $138.8 billion in 1982 (Appendix 1.8). But, even though Japan has become a world leader in exports, this role has been achieved at a very high cost, as 1979 recorded a negative balance of trade of $7.6 billion, and another deficit of $10.7 billion took place in 1980. However, there has been a favorable balance since then reaching $44.4 billion in 1984. These data show that, while Japan has on balance had a positive balance of payments situation over the past two decades, foreign exchange earnings can be a problem. The exchange rate from yen to dollars is given in Appendix 1.9.

Exports of agricultural commodities only account for about five tenths of one percent of all Japan's exports. But, imports of agricultural commodities make up 13-14 percent of all imports (Appendix 1.10). While the proportion of imports that are agricultural has continuously declined, in absolute terms agricultural imports grew 2.5 times from 1970-1984 on a total value basis.

Growth in agricultural imports has taken place fairly evenly across all agricultural commodities on a percentage basis although, on a total basis the major commodities were cereals (Appendix 1.11). Imports of live animals on a total value basis have remained modest, about 14-18 billion yen since 1980. Meat and meat preparations have increased seven times since 1970 now reaching 437 billion yen. Dairy product and egg imports have tripled in that time, reaching 72 billion yen in 1984 (Appendix 1.12). They have declined since 1981.

Beef imports remained at a very modest level through 1969 at which time they only amounted to about 18,000 metric tons (Appendix 1.13). Then, in 1970, they began climbing rapidly, reaching 170,000 tons in 1973. Severe restrictions were placed on imports and they dropped to 40,000 tons in 1974, but have continued to climb since then, reaching 189,000 tons in 1979 and 208,000 tons in Japan's 1983 fiscal year (April-March).

Pork imports followed a similar pattern to beef but, after reaching a high of 208,000 tons in 1975, declined and only passed the 200,000 ton level again by 1980 (Appendix 1.14). They reached 271,000 tons in 1983. Chicken meat imports have continued to grow steadily but are a very small part of domestic production. For example, domestic production was 505,000 tons in 1970 while imports were just 12,000 tons or 2 percent of total consumption (Appendix 1.15). By 1983 imports had grown to 100,000 tons but that is less than 1 percent of domestic production which had climbed to 1.3 million tons.

The United States is Japan's major trading partner. However, although the U.S. takes one fourth of Japan's exports, the U.S. only provides about 17-18 percent of Japan's imports. But, these overall data somewhat mask the importance of Japan to the U.S. as a trading partner, especially in terms of agricultural trade. Japan now exceeds the EEC as an importer of U.S. agricultural commodities. For example, Japan imports 81 percent of the United States' exports of lemons and limes, 33 percent of its grain sorghum exports, 36 percent of its pork exports, and 71 percent of beef exports (Table 1.3).

SELF-SUFFICIENCY

A major source of concern to the Japanese is the extent to which the country is able to produce sufficient foodstuffs. Many still remember the hardships and hunger during and after World War II, when there were shortages of almost all agricultural commodities. Furthermore, because Japan is so heavily dependent on trade for its livelihood, there is widespread recognition of the immediate impact from shipping strikes, war

Table 1.3. Japan's share of major U.S. farm exports, 1984

	Proportion of total U.S. exports to Japan	Japan's rank as a recipient of U.S. exports
	- Percent -	
Lemons and limes	81	1st
Beef and veal	71	1st
Grapefruit	58	1st
Pork	36	1st
Grain sorghum	33	1st
Corn	28	1st
Feed grains	28	1st
Hides and skins	24	1st
Poultry meats	24	1st
Oranges and tangerines	23	3rd
Soybeans	22	1st
Grains and feeds	19	1st
Wheat	08	2nd

Source: Calculated from U.S. Dept. of Agr., FATUS, January/February, 1985.

and whims of trading partners. For example, one of the most damaging agricultural policies of the United States vis-a-vis Japan was the embargo on shipments of soybeans and soybean products to Japan in 1973 as a result of a temporary decline in U.S. production.

The concern about foodstuffs self-sufficiency is partly manifested in data which show that agriculture has continuously declined in importance as Japan has industrialized. In 1970 agriculture, forestry and fisheries contributed to about 8 percent of gross domestic product (GDP); by 1983 it had declined to just 3 percent of GDP. Japan has reason to be concerned about food production since their self-sufficiency ratio in food has continuously declined, from 90 percent in 1960 to 83 percent in 1965, down again to 78 percent in 1970, then to 72 percent in 1980 a level at which it appears to have stabilized (Table 1.4).

All commodities have registered a decline in the ratio, but the greatest impact has been in cereals,

Table 1.4. Self-sufficiency ratios in food commodities, Japan, 1960-1983

Item	Year							
	1960	1965	1970	1975	1980	1981	1982	1983
	Percent							
Total food	90	83	78	74	72	72	72	71
Cereals, including feed grains	82	61	45	40	33	33	33	32
Cereals as a principal food								
Total	89	80	74	69	69	69	69	69
Rice	102	95	106	110	87	92	93	94
Wheat	39	28	9	4	10	10	12	11
Soybeans	28	11	4	4	4	5	5	4
Vegetables	100	100	99	99	97	97	98	96
Fruits and nuts	100	90	84	84	81	77	79	81
Meat								
Total	91	90	89	77	81	80	80	80
Beef	96	95	90	81	72	75	71	70
Pork	96	100	98	86	87	86	87	85
Hen eggs	101	100	99	97	98	97	98	98
Milk and milk products	89	86	89	82	86	84	85	86
Sugar	18	30	23	16	29	28	31	30

Source: MAFF. Abstract of Statistics on Agriculture, Forestry, and Fisheries, 1984. Tokyo, March 1985.

Note: $\text{Self-sufficiency ratio} = \frac{\text{Value of domestic production}}{\text{Value of domestic consumption}} \times 100$

including cereal for feed in which domestic production dropped from 82 percent of total usage in 1960 to about 33 percent since 1980. The ratio for soybeans, a commodity of major importance in the Japanese diet, declined from 28 percent in 1960 to 4-5 percent since 1970. The self-sufficiency ratio for beef dropped from 96 percent in 1960 to 72 percent in 1980, where it has about stablized. Pork went from 96 percent to about 86-87 percent over the same 20 year period. Eggs, a highly perishable commodity, have remained near 100 percent, and milk and milk products have stayed in the high 80's.

SUMMARY

Japan is characterized by an extremely heavy population density which on a per unit of arable land basis is about double that of China and more than 20 times that of the United States. The land is very intensively cultivated and yields are quite high. Despite a slowing down in the population growth rate, total population has expanded from 98.5 million people in 1965 to 119 million in 1984. It is expected to reach 123 million in 1990, 128 million in the year 2000, and 130 million by 2010. Only 13.1 percent of Japan's land is now arable and each year the total is reduced due to urban sprawl.

The persistent downward trend in the food self-sufficiency ratio, 90 percent in 1960 to 72-73 percent since 1980 is viewed with great alarm and is a major factor contributing to Japan's reluctance to liberalize imports of meat, citrus and other agricultural commodities. Another factor is a concern about westernization of dietary habits. The impact on meat consumption, and beef in particular, is examined in the next chapter.

2 Dietary Patterns and Beef Consumption

Food preferences and consumption patterns depend on tradition, climate, resource availability and propensity to trade, just to mention a few factors. The review of Japan's agricultural economy provided in the previous chapter points to radical changes which have taken place in the past several decades and the last one in particular. Those changes, combined with equally impressive ones in the urban sector, have altered economic patterns, cultural habits--and diet. The objective of this chapter is to provide an explanation of historical and current forces which will influence Japanese demand for beef in the next quarter century. The results from this analysis, along with supply side information developed in the next several chapters, provide the basis for long-term analyses developed later.

TRADITIONAL BEEF CUISINE IN JAPAN

The Beginning of Meat Eating

The topic of beef consumption in Japan begins with revocation of the prohibition against meat eating in 1882 just after the Meiji Restoration. The eating of beef, pork and other animal meats had been prohibited for more than a thousand years until the Meiji Restoration, mainly because of the Buddhist dogma that killing and eating of four-legged animals was cruel. Curiously enough, fish and birds were excluded from this dietary ban.

The deficiency in protein intake due to this dietary ban was compensated for by eating fish, processed food from soybeans, and rice. Fish, the main source of non-vegetable protein intake, was supplied in abundance by the surrounding seas. It is remarkable that this dietary pattern has continued down to the present, as the Japanese people still take a large part of their protein from fish and soybeans. When the ban on meat eating was removed a new beef dish, sukiyaki, was developed. It is described on a menu of a Japanese restaurant in the U.S., as "various vegetables and paper-thin sliced prime sirloin cooked in a special soy-sauce."[1]

Generally speaking, high quality beef tastes best when it is cooked by grilling, broiling, roasting or frying, i.e. dry heat methods. Cooking beef with moist heat, i.e. by boiling, simmering etc., is best to enhance the taste of lower quality beef. Stewed beef is a good example of the latter type of cuisine.[2] Sukiyaki is an exception to these generalizations for, in this dish, the highest quality most marbled beef is "boiled" in a pot. The other main feature in sukiyaki preparation is the use of "prime sirloin" sliced like wafers or paper. There are two questions. One is why the Japanese people developed such a unique beef dish. The other is how sukiyaki became world-famous in light of its exceptional cooking method. The answer is that until the Meiji Restoration, Japanese maintained the dietary ban on meat not only because they had believed Buddhist dogma, but also due to the influence of Shinto, the Japanese native religion.

The combinations of religions over the long history of the Japanese people made such a deep impression that they came to dread and hate the sight of blood and dead bodies of both human beings and animals. Therefore, when the dietary ban was removed after the Meiji Restoration, Japanese would not eat meat in a form which tended to show blood. They asked butchers to cut their beef into paper-thin slices to avoid seeing blood in the kitchen or on the table. In addition, they would dip the sliced beef into boiled water according to the Shinto dogma which held that water was a means of purification.

Sukiyaki and Overestimation of the Marbledness of Beef

The boiling of high quality beef, especially paper-thin sliced beef, makes it tough though leaving a nice soup. The question then is how Sukiyaki could become such a delicacy. Two or three factors should be considered. First, Japan has had a native breed of beef cattle, the Wagyu (Japanese Black), which has the ability to produce well-marbled beef. In fact, some loin cuts are so well-marbled that Japanese call them "shimofuriniku" (frost-like beef). Such highly marbled beef does not become tough when cooked with moist heat. At high temperature the fat in the muscle melts and covers the outside of the beef, minimizing the loss of natural juices and flavoring the beef from the inside.

The second factor is that Japan has a rich variety of vegetables in all four seasons as well as foods processed from soybeans such as shoyu (soysauce) and tofu (soybean curd). For Japanese, who enjoy food not only by eating but also visually, beautifully marbled beef, white tofu, and various green and white vegetables arranged in a pot are very pleasing. This mixture, when boiled together with sugar and soysauce is sukiyaki, and its taste greatly depends on marbledness.

Returning to historical development, it must be recognized, of course, that not all Wagyu produce high quality meat, and even the best quality Wagyu have inferior parts. Japanese have utilized this lower quality beef in a simple dish, nimono, as a kind of seasoning. Various vegetables are boiled with soysauce and a small amount of beef in the form of paper-thin slices. It is also true in this case that the more the beef is marbled, the more tasty the dish. Thus, until a few decades ago, almost all beef was retailed in the form of paper-thin slices, and marbled beef was valued higher. But this situation raised the problem of overestimation of marbled beef.

Moderately rather than highly marbled beef is enough to make Nimono or even sukiyaki tasty for a daily diet. But even nowadays most Japanese wholesalers and retailers, and sometimes even consumers, are apt to think that the more beef is marbled, the more expensive it should be. Though the overestimation of marbled beef has its root in historical dietary patterns, it seems to retard rationalization of the beef production and marketing system.

MEAT CONSUMPTION AFTER WORLD WAR II: A BIG INCREASE

The Japanese Dietary Pattern before World War II

As Table 2.1 shows, Japanese took more than 60 percent of their total energy intake from rice in the pre-war days. Most vegetables were eaten as salty pickles or with misosoup. The fish were consumed mostly in the form of dried fish with salt. These side dishes stimulated people to take more rice.

Rice has unique characteristics as a staple food. It contains most of the necessary nutrients, and boiled rice is easily taken in large quantities by itself or with a small salty side dish. This dietary pattern, i.e. taking most of the necessary calories from rice, was low cost compared with a diet of bread supplemented with butter or other livestock products. In the pre-war days, the Engel ratio of the Japanese people was

Table 2.1. Daily per capita caloric intake, Japan, 1930-1983

Year	Rice	Meat	Marine products	Fats & oils	Other	Total
			Percent			
1930	61	<1	03	01	35	100
1935	59	<1	03	01	37	100
1939	63	<1	03	02	32	100
1946	60	<1	02	<1	38	100
1950	53	<1	04	01	42	100
1955	48	01	04	03	44	100
1960	48	01	04	04	42	100
1965	44	02	04	07	43	100
1970	36	03	04	09	48	100
1975	34	04	05	11	46	100
1980	30	05	05	14	46	100
1983	28	06	05	15	46	100

Source: Appendix 2.1.

far lower than those of nations which had nearly the same per capita national income as Japan. Even in the mid-1960's, when the Japanese per capita national income had reached about 700 U.S. dollars, Japan's Engel ratio was about 35 percent which was nearly equal to that of nations with per capita national incomes of 1,500-2,000 U.S. dollars. This low Engel ratio was a function of the less expensive Japanese dietary pattern.

Westernization of the Japanese Diet

Japan's rapid economic growth, which began in the mid-1950's and lasted until the late-1970's, changed the dietary pattern of the Japanese people completely. The traditional diet which was characterized by rice, vegetables and fish gave way to a new makeup, the features of which might be summarized as follows: (1) a decrease in the per capita consumption of rice and an increase in that of flour-based foods, (2) a rapid increase in the consumption of livestock products, (3) an increase in the consumption of fruits and vegetables as salad and dessert, and (4) a decrease in the use of traditional seasoning, like miso and shoyu, and an increase in the use of sugar, oil, mayonnaise, etc.

The change in Japan was called the modernization or the westernization of the dietary pattern. The most important factor which caused this change was, of course, the increase in per capita disposable income. But social factors should not be overlooked. For example, during the period of starvation immediately after World War II, the Japanese people were obliged to become familiar with foods based on wheat or corn. The school lunch program introduced by the Occupation Authority also accelerated the diversification of staple foods. After the rapid economic growth period started, important factors in the dietary change were: (1) the expansion of cities, (2) the population shift from rural areas, (3) the increase in the number of nuclear families, and (4) the increase in the number of working housewives.[3]

Increase in the Consumption of Livestock Products

By the mid-1950's Japanese livestock production had recovered from the great damage inflicted by the deficiency of feed during the war and started on a path of rapid growth which began with expansion of the poultry and hog sectors. This was partly because Japan was so mountainous that the limited flat land could not be used as pasture or in forage production for cattle. Another reason was demand-related.

During the 1960's the growth rate of Japanese GNP was very high, but the increase in per capita disposable income was not sufficient for the Japanese people to enjoy expensive livestock products. Consequently, the increase in disposable income was spent on poultry and pork products for which production costs were low. As Table 2.2 shows, during the period of rapid economic growth chicken production grew the fastest followed by pork production, while the increase in the rate of beef production was the lowest.

There was another factor particular to Japan regarding the low rate of beef production increase. Until the late-1950's almost all farmers in Japan kept one head of cattle or one horse as a draft animal. The peak of cattle inventory (Wagyu) was about 2.7 million head in 1956 kept on 2.3 million farms (there were 6.0 total farms in 1955). But the mechanization of agriculture and the increase in the number of part-time farmers, both of which started in the mid-1950's, caused a rapid decrease in the number of cattle kept on farms. As the Japanese government could not find any

Table 2.2. Expansion of Japan's livestock production (domestic) during the postwar period of rapid economic growth, 1960-1975.

Year	Milk	Eggs	Pork	Chicken	Beef
	- - 1,000 MT - -		1,000 MT (carcass wt)		
1960	1,939	696	149	103	141
1965	3,241	1,330	431	238	196
1970	4,489	1,766	779	496	282
1975	5,008	1,807	1,023	459	335
1980	6,498	1,992	1,430	1,120	431
1983	4,088	2,091	1,433	1,257	505

Source: MAFF, Food Balance Sheet, various years.

effective policy to increase the number of beef cattle in place of draft cattle in the early-1960's, the short supply of beef brought about a rise in its price. Chicken or pork was then substituted for beef.

Even during the 1960's, though consumption of livestock products increased greatly, most of the increase was in the form of substitutes or imitations of pure and expensive products. This was especially true for the processed foods from livestock products. In those days, Japanese consumed much compound milk (compounded from imported skim milk and butter), pressed ham (an imitation of ham made from small pressed pieces of imported mutton or horse meat) and other imitation products. But, in the late 1960's they began to want "true and pure" livestock products; for example, true whole milk, true ham from pork, natural cheese instead of processed cheese, and beef instead of chicken and pork.

Fattening of dairy steers began in the late 1960's. Prior to that time most of them had been killed as vealers, mainly for sausage. Then, owing to the increased supply of domestic fed dairy steers and of imported beef from Oceania, per capita consumption of beef began to increase in the late 1960's (Table 2.3). At the same time, numbers of Wagyu stopped decreasing although inventories have not increased much today. As a result, even though Japanese still mainly consume chicken and pork, beef consumption began to increase gradually. By the late 1970's total beef consumption exceeded 500 thousand metric tons, divided evenly among imported beef, domestic Wagyu beef, fed dairy steers, and culled dairy cows. Per capita beef consumption has continued to increase.

Table 2.3. Per capita consumption of beef, pork, poultry and milk products in Japan, 1960-1984

Fiscal year	Beef and veal[a]	Pork	Chicken	Total three meats	Milk products
	- - - - - - - - - - - - kg - - - - - - - - - -				
1960	1.1	1.1	0.8	3.0	22.2
1965	1.5	3.0	1.9	6.4	37.5
1970	2.1	5.3	3.7	11.1	50.1
1975	2.5	7.3	5.3	15.1	53.3
1980	3.5	9.6	7.7	20.8	62.1
1984	4.2	9.6	8.6	22.4	67.1

Source: MAFF. <u>Food Balance Sheet, 1983</u>. Tokyo, March, 1985.

[a]Retail weight basis.

A Turning Point in Modernization of the Diet

As just described, Japan experienced a dietary change in the period of rapid economic growth. Then came the oil crisis of 1973-1974 which was a turning point for the Japanese economy as it entered into a new stage of stable or low growth. At the same time, calorie intake reached its peak and became stagnant (Table 2.4). From 1950 per capita rice consumption began a steady decrease which has continued up to now (Table 2.1). In the middle of the 1970's per capita consumption of fruits and vegetables peaked out as did flour products, eggs and marine products. Per capita consumption of meat, milk and milk products increased from 1975 until 1980, but the slope of their trend (including that of beef) became less steep, especially after the 2nd oil crisis of 1978-1979.

It is worth noticing that in 1975 protein intake from livestock products (meat, eggs, milk and milk products) exceeded that from marine products. This is remarkable not only because livestock products had become the principal source of protein intake, but also because, with stagnant growth in food consumption, they could exceed marine products as a source of protein intake.

Table 2.4. Supply of protein, fat and calories per capita in Japan, 1960–1983

Fiscal Year	Protein, Livestock products				Protein fish and shellfish	Total protein intake from all sources	Total fat intake	Total caloric intake	Ratio of protein (P), fat (F) and carbohydrate (C) in total caloric intake		
	Meat	Eggs	Dairy products	Subtotal					P	F	C
	Grams							Calories	Percent		
1960	2.8	2.2	1.8	6.8	14.6	68.9	29.1	2537	12.1	11.5	76.4
1965	4.9	3.8	3.0	11.7	14.2	94.4	44.9	2542	12.2	15.1	72.7
1970	6.8	4.9	4.0	15.7	15.2	77.4	57.2	2586	12.3	19.1	68.6
1975	9.0	4.6	4.2	17.8	17.1	79.6	65.1	2404	12.8	22.0	65.2
1980	11.3	4.8	4.9	21.0	17.8	81.8	75.5	2504	12.8	25.1	62.1
1983	11.9	4.9	5.3	22.1	17.4	81.6	79.9	2593			

Source: Derived from MAFF, Food Balance Sheet, 1983. Tokyo, March, 1985.

DIVERSIFICATION OF BEEF COOKING METHODS

Cooking Methods of Pork and Chicken in Japan

As mentioned earlier, the traditional beef cuisine in Japan has been sukiyaki which, when cooked in this way with highly marbled beef has an exquisite taste. "Shabu-shabu" should also be mentioned as belonging to this class of beef cuisine. Shabu-shabu is a beef dish in which one dips a piece of paper-thin sliced and highly marbled beef into a pot of boiling water for an instant and then eats it. This dish, similar to the "Meat Fondue" of southern Europe, became rather popular in Japan after World War II. It also tastes wonderful when the beef is highly marbled.

Beef consumption increases since the late-1960's have largely been of domestic dairy steer beef and imported grass-fed beef. This change in beef consumption could not help but alter the mix of beef cooking techniques. But, before taking up this problem, let us examine the traditional methods of cooking pork and chicken.

Japanese have eaten pork cooked in various ways from the earliest times that pork consumption became popular, i.e. just after the Meiji Restoration. Some of the pork recipes devised as side dishes to boiled rice were particular to Japan. Other pork dishes were similar to those of Americans, who typically cook pork by broiling, roasting or frying. As for chicken, the Japanese people consumed the least quantity and paid the highest price among meats through the 1950's (Table 2.5). In those days most chicken was consumed by boiling, much like sukiyaki or shabu-shabu. After World War II the U.S. occupying army introduced roast or broiled chicken which at first was very strange to Japanese. As chicken consumption began to increase in the 1960's, much of the initial increase was brought about by combining it with traditional cuisine. But consumption of roast and broiled chicken began to increase and gradually occupied a larger share. Cooking methods for pork and chicken have been more diverse in Japan than those of beef. Accordingly, they have played a more important role in the daily diet.

Diversification in Beef Cooking Methods

The increased supply of domestic dairy steer beef and imported grass-fed beef was accompanied and supported by the diversification of beef cooking methods. One reason has been the influence of the preceding diversification of pork and chicken cooking methods.

Table 2.5. Price and per capita consumption comparison between beef, pork and chicken, by households in Japan, 1955-1983

Year	Beef	Pork	Chicken	Proportion of beef: Pork	Proportion of beef: Chicken
			Average retail price[a]		
	- - - Yen/100 g - - - -			- - Percent - -	
1955	37.8	41.8	42.0	111	111
1960	50.3	50.1	51.5	100	102
1965	75.0	66.1	60.1	88	80
1970	123.5	82.7	65.4	67	53
1975	242.2	139.3	103.9	58	43
1980	312.4	138.6	99.8	44	32
1983	318.3	154.5	102.8	48	32
			Average annual quantity[a]		
	- - - - - - kg - - - - -			- - Percent - -	
1955	7.93	3.14	1.14	40	14
1960	9.47	4.44	1.98	47	21
1965	9.63	9.83	5.16	102	54
1970	7.51	14.75	9.04	196	120
1975	8.21	18.89	11.29	230	138
1980	9.64	21.36	14.44	222	150
1983	10.21	19.32	14.38	189	141

Source: Derived from PMO, Statistical Bureau, Annual Report on the Family Income and Expenditure Survey (all households in the cities with population of 50,000 or more).

[a]"annual quantity" is the average quantity purchased by households during the year; "average price" is the average price households paid during the year.

Gradually, Japanese began to enjoy beef cooked by the same method as pork and chicken. This trend was accelerated by the introduction and extension of cooking facilities such as grills, ovens and ranges in households from the latter half of the period of rapid economic growth. For example, hamburgers and other ground beef dishes became popular in those days, following the acceptance of ground pork cuisine. (In Japan, the mixture of ground beef and ground pork is viewed more favorably than ground beef by itself.) The diversification in beef cooking methods included not only these beef dishes which were cooked with dry heat, but also those with moist heat, such as stewed beef and curried rice with beef.

There was another and more important factor. It was the introduction and extension of "yakiniku" immediately after World War II, a kind of meat dish, the special feature of which is to grill or barbeque (indoors) small pieces of meat (not so thin as a wafer) with heavy sauce containing red pepper. This meat dish is supposed to have been introduced from the Korean Peninsula during or after World War II. Also, at this time, cattle and hog by-products were first grilled. Further, during this post-war period small slices of pork and chicken were broiled on iron plates with a heavy sauce and, when Japan started importing grass-fed beef, most of this was consumed in dishes of this type. Some domestic dairy beef is also utilized in this way.

The forms of beef retail cuts have become manifold with the diversification of beef cuisine. In the days when most beef was consumed as sukiyaki or something similar, one found only paper-thin sliced beef in the show windows of meat retailers. More recently it is common to see not only paper-thin sliced beef but also yakiniku sliced, steak sliced, cube cut, etc. We now examine the current situation of beef cuisine and retail cuts as derived from consumer surveys.

Current Situation of Beef Cuisine

Two surveys about beef consumption patterns will be examined. One is the "Survey on the Tendency of Seasonal Meat Consumption" by the Research Center for Food Demand and Supply (in Tokyo), a group that has annually interviewed 2,000 households from 19 cities all over Japan. (Hereafter we call this the "Tendency Survey".) Another is the "Comparative Survey on the Beef Consumption Customs between the U.S. and Japan" which was performed by T. Yoshida with the assistance of J. Simpson, Prof. Kokoris and others.[4] In this survey, 59 parents of elementary school children in Gainesville, Florida and 80 female students of Northeastern Illinois University in Chicago, Illinois were sampled. In Japan, 84 parents of students from one elementary school and one kindergarten in Kyoto, and 29 female students of Kyoto Prefectural University were sampled. (Hereafter we will refer to this as the "Comparative Survey" for brevity.)

First, by examining the "Comparative Survey," the main beef cooking methods in the households of the U.S. and Japan can be compared. Table 2.6 shows that in the U.S., except for hamburger (81.2 percent) the percentages of beef steak, beef roast, beef pot roast, barbecue and stewed beef range from about 30 to 40 percent while in Japan the beef cooking methods tend to concentrate in 3 types, for the percentages of sukiyaki, yakiniku and hamburger exceed 60 percent. (Though the percentage of curried rice is high, only small amounts of beef are used in this dish as it is a kind of seasoning.) In summary, beef dishes in the U.S. are cooked mostly with dry heat and have variations among broiling, roasting or grilling. Japan, on the other hand has 3 dominant cooking methods.

Table 2.6. Beef preparation methods in Japan and the United States, Comparative Survey[a]

Preparation methods	Japan		United States	
	Responses	Percentage	Responses	Percentage
Stew	41	36.3	41	29.7
Steak	11	9.7	64	46.4
Roast	3	2.7	63	45.7
Hamburger	76	67.3	112	81.2
Pot roast	-	-	47	34.1
Barbecue	-	-	43	31.2
Shabu-shabu	11	9.7	-	-
Sukiyaki	70	61.9	-	-
Yakiniku	68	60.2	-	-
Beef curry rice	63	55.8	-	-
Others	6	5.3	20	14.5
Total possible responses	113	-	133	-

[a]Survey conducted in 1983. See footnote 4. Also, the percentages do not equal 100 because the respondents were asked to check the most frequent 3 methods from the list.

The fact that in Japan beef dishes have been diversified from sukiyaki to those cooked with dry heat is supported by results of the "Tendency Survey" (Table 2.7). Trends cannot be determined from these data since they only cover two years. However, they do show that in winter (December and March) the percentage of beef dishes with moist heat increases, and in summer

Table 2.7. Seasonal variation in Japanese beef preparation methods

Method	Fall Sept 1982	Fall Oct 1982	Fall Sept 1983
	- - - - -	Percent	- - - -
Grilling broiling roasting, etc.	37.2	32.2	41.7
Broiling, simmering, etc.	46.5	52.2	43.8
Frying	11.8	10.8	11.8
Others and no reply	4.4	4.8	2.7
Total	100.0	100.0	100.0

	Winter Dec 1983	Spring March 1984	Summer June 1983
	- - - - -	Percent	- - - -
Grilling broiling roasting, etc.	32.8	32.2	40.7
Broiling, simmering, etc.	53.9	54.2	42.1
Frying	9.3	10.0	14.0
Others and no reply	4.0	3.7	3.2
Total	100.0	100.0	100.0

Source: Research Center for Food Demand and Supply. "Survey on the Tendency of Seasonal Meat Consumption." Tokyo, 1984.

they decrease. In the winter dishes such as sukiyaki and shubu-shabu are well-received due to their cooking method. In summer (e.g. in June and September of 1983), the percentage of beef dishes cooked by grilling and the like almost exceeded that of boiling and similar methods. It might be thought that in summer the frequency of beef dishes with dry heat would increase beyond that of dishes with moist heat. In fact, the former might be nearly equal to the latter for the whole year.

Table 2.8, which shows the change in the type of retail beef cuts purchased by consumers from 1979 to 1983 indicates that about 50 to 60 percent of beef has

Table 2.8. Japanese beef purchases by retail cut, 1979-1983

Year and month	Beef Cut[a]			
	Paper-thin slice	Yakiniku slice	Steak	Cubed
1979 (Oct)	67.2	23.2	13.7	11.2
1980 (Oct)	55.8	32.2	12.5	13.1
1981 (Oct)	61.5	33.9	16.8	17.5
1982 (Sept)	55.9	36.9	16.1	14.4
1983 (Sept)	53.5	35.3	15.5	15.8

Year and month	Beef Cut[a]			
	Joint	Chopped	Ground	Other
1979 (Oct)	6.7	4.0	14.3	0
1980 (Oct)	4.5	5.3	15.8	0
1981 (Oct)	3.9	6.4	37.1	0
1982 (Sept)	3.6	6.0	16.1	1.1
1983 (Sept)	4.0	6.2	16.9	0.3

Source: Research Center for Food Demand and Supply. "Survey on the Tendency of Seasonal Meat Consumption". Tokyo, 1984.

[a]The interviewees checked 2 or less types of the beef retail cut which they purchased during a certain week of each month. The percentages are the ratio of the numbers checked to the numbers of total interviewees. Thus, each line sums to more than 100.

been retailed in the form of paper-thin slices. The yakiniku slice, ground beef, steak slice and cube cut followed sucessively. It appears that paper-thin slices are becoming less important percentage-wise being replaced by yakiniku slice, steak slice, cube cut and ground beef. This is obviously the result of diversification in beef dishes. Overall, it is clear that Japanese beef cuisine has broadened from traditional styles cooked with moist heat to those cooked with dry heat.

A COMPARISON OF CONSUMERS' IMAGE OF BEEF BETWEEN JAPAN AND THE U.S.

A Comparison of Meat Consumption from the "Comparative Survey"

First, as can be seen from Table 2.9 there is a remarkable difference in distribution of meat consump-

Table 2.9. Meat consumption patterns per household by large or heavy users, Japan and the United States[a]

Consumption patterns	Japan	United States
	- - - - Percent	- - - -
Beef (a)	17.0	37.9
Beef and pork (b)	10.4	6.1
Beef, pork and chicken (c)	20.8	5.3
Beef and chicken (d)	13.2	23.5
Pork (e)	3.8	1.5
Pork and chicken (f)	10.4	0.0
Chicken (g)	8.5	9.8
Subtotal	84.1	84.1
Small amount of all meat (h)	16.0	15.9
Total	100.0	100.0

[a]Large or heavy consumption is "almost every day" or "once every two or three days". Small is "once a week" or "once or twice a month." Comparative survey carried out in 1983.

tion types between the U.S. and Japan for, in the U.S., the greatest concentration (61 percent) is in beef only, and beef and chicken. In Japan on the other hand, meat consumption is scattered among several types which are combinations of beef, pork and chicken. This study shows that in the U.S. beef and/or chicken is eaten almost every day, while the usual Japanese meat consumption pattern is a combination of each kind of meat.

Next is a comparison of the change in beef consumption of both countries by meat consumption type. In response to the question "do you think your beef consumption has increased or decreased this year compared to that of 4 or 5 years ago?" about 46 percent of the U.S. respondents replied that their beef consumption had decreased, while the reply of about 48 percent of the Japanese respondents was "increase" (Table 2.10). These survey results show that in the U.S. the biggest decrease is in type (2) (the type consuming a large amount of pork and/or chicken), and the second biggest decrease in type (3) (the type consuming a small amount of all meat). In Japan, the biggest increase can be seen in type (1) (the type consuming a large amount of beef, or of beef and pork and/or chicken), the least increase in the type (2), the biggest decrease in the type (3), and the least decrease in the type (2).

The overall findings are that in Japan, though the increase in beef consumption made the number of type (1) households increase, in the existing households of the types (2) and (3) there is no big increase in beef consumption. The existing households of the types (2) and (3) are stable in their meat consumption. On the other hand, in the U.S., the decrease in beef consumption which has been increasing in the number of households of the types (2) and (3) is still diminishing beef consumption in the type (1) households.

A Comparison of Consumer's Beef Image by the Semantic Differential Technique.

Two methods were used to determine consumer's image of beef: the semantic differential technique (the SD technique in short) and the rating scale method (the RS method in short). In the SD technique, the respondents were asked to choose one among 1-6 scales which were located between an adjective and its antonym modifying beef. In the RS method, they were asked to choose one from 5 rating scales concerning some opinion about beef.

Table 2.11 shows the results from the SD technique. To the items of the first group (those relating

Table 2.10. Change in beef consumption patterns, Japan and the U.S.

Meat consumption type[a]	Japan			
	Increased	Unchanged	Decreased	Total
Large amounts of beef (and pork and/or chicken) (1)	37 6 (57.8)	21 (9.4)	64 (32.8)	(100.0)
Large amounts of pork and/or chicken (2)	7 (29.2)	12 (50.0)	5 (20.8)	24 (100.0)
Small amounts of all meats (3)	6 (35.3)	4 (23.5)	7 (41.1)	17 (100.0)
Total	50 22 (47.6)	33 (21.0)	105 (31.4)	(100.0)

Meat consumption type[a]	United States			
	Increased	Unchanged	Decreased	Total
Large amounts of beef (and pork and/or chicken) (1)	32 25 (33.3)	39 (26.0)	96 (40.6)	(100.0)
Large amounts of pork and/or chicken (2)	3 (20.0)	1 (6.7)	11 (73.3)	15 (100.0)
Small amounts of all meats (3)	2 (9.5)	8 (38.1)	11 (52.4)	21 (100.0)
Total	37 34 (28.0)	61 (25.8)	132 (46.2)	(100.0)

Source: "Comparative Survey."

[a]The question was "Do you think your beef consumption in pounds has increased or decreased this year compared to that of 4 or 5 years ago?" The meat consumption types were defined as (1)=(a)+(b)+(c)+(d) (of the Table 2.9), (2)=(e)+(f)+(g)(ditto) and (3)=(h) (ditto).

Table 2.11. Comparative analysis of consumer's image of beef in Japan and the United States using the semantic differential (SD) technique

Factor group	Item	Positive attribute	Average ranking[b] 1 2 3 4 5 6	Negative attribute	Average ranking Japan	Average ranking U.S.	Variance Japan	Variance U.S.	Significance level[a] Average ranking	Significance level[a] variance
1.	Sensory, i.e. taste, flavor, visual, etc.	Flavorful		Flavorless	3.4	2.3	0.972	1.152	***	
		Tasteful		Tasteless	2.5	2.1	1.232	1.134	***	
		Tender		Tough	2.8	2.7	0.760	1.187		***
		Beautiful		Ugly	3.2	3.5	1.013	1.785	**	***
		Clean		Dirty	3.1	2.8	0.865	1.350	**	***
2.	Health, nutrition, etc.	Healthy		Unhealthy	2.9	2.9	0.849	1.549		***
		Nutritious		Unnutritious	2.4	2.4	0.742	1.491		***
		Fatty		Lean	3.2	3.2	1.120	1.737		***
3.	Economy	Gorgeous		Plain	2.9	3.8	0.767	1.815	***	***
		Expensive		Inexpensive	2.4	2.3	1.338	0.952		**
4.	Role and importance in daily dietary life	Necessary	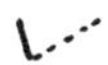	Unnecessary	2.5	3.7	1.543	2.041	***	*
		Want to eat		Don't want to eat	2.6	2.6	1.311	1.622		

Source: "Comparative Survey"

[a]***, **, * denotes that t-test of difference is significant with the probability of 0.01, 0.05 and 0.1.

to taste, flavor and the like), the U.S. consumer's estimations are higher than those of the Japanese consumers except on the item "beautiful-ugly." Americans seem to take more pleasure in tasting beef than Japanese. It is interesting that Japanese enjoy beef not only with their mouths but also with their eyes, for they feel beauty in seeing beef. (But they are not so particular about the flavor of beef as are Americans.)

Consumer attitudes in both countries seem to be equivalent since, for the items concerning health and nutrition, Japanese have the same feeling about the merits and demerits of beef to human health as do Americans. To the items, "gorgeous-plain" and "necessary-unnecessary," the Japanese respondents gave lower marks. This seems to show that Japanese want to eat more beef but they can't help giving it up as they consider it gorgeous. But, consumers' responses to the items, "expensive-inexpensive" and "want to eat-don't want to eat" are similar in the two countries.

Now, turn to the analysis by the RS method (Table 2.12). First, the items concerning the favorableness or unfavorableness of beef to human health. The U.S. consumers rate the healthiness of beef highly and at the same time they regard beef as unhealthy food. Japanese consumers also consider beef to be both favorable and unfavorable to human health, but the intensity of opinion is moderate compared to that of the Americans, and to some of the items they responded neutrally. However, some attention should be paid to the Japanese consumers showing stronger agreement with the opinion, "Beef is full of cholesterol", and "We should not eat beef so often" than the U.S. consumers. This finding seems to show that Japanese consumers, who eat only one-eighth or ninth of Americans per capita beef consumption, are catching up with the U.S. consumers in their anxiety about the hazards of overeating beef.

Next, the items which relate to the taste of beef. The U.S. consumers agree strongly with the opinion, "Beef tastes good," but they are neutral about "Beef is the best tasting meat." This seems to be a reflection of the U.S. consumer's movement from massive consumption of beef to consumption of more chicken and pork. The Japanese consumer's rankings on taste are moderate compared with those of U.S. consumers.

As for the opinions on economy, both the U.S. and Japanese consumers agree with the opinion, "Beef is expensive" to the same degree. The U.S. consumers consider beef to be expensive to the same degree as do Japanese consumers who face a price 2 or 3 times higher than that in the U.S. for many beef cuts.

Japanese consumers show stronger agreement than Americans to most of the items relating to the role and importance of beef in daily diet and the future intention of consuming beef. They seem to rate the role of

Table 2.12. Comparative analysis of consumer's image of beef in Japan and the United States using the rating scale (RS) method

Item and attribute	Average ranking[a] 1 2 3 4 5 6	Average ranking Japan	Average ranking U.S.	Variance Japan	Variance U.S.	Significance level[b] Average ranking	Significance level[b] variance
(1) Healthfulness							
Beef is rich in vitamins		3.2	2.7	0.650	0.871	***	*
Beef is good for health		2.7	2.6	0.726	1.146		***
Beef is rich in nutrition		2.2	2.1	0.708	0.833		
(2) Unhealthfullness							
We should eat leaner beef		2.8	2.0	0.929	1.131	***	
Beef tends to cause heart attacks		3.4	2.9	0.925	1.411	***	**
Beef is rich in fat		2.9	2.6	0.795	1.521	**	***
Consumption of beef causes high blood pressure		3.2	2.9	1.046	1.149	**	
Beef makes people fat		3.1	2.9	0.973	1.533	*	
Beef is full of cholesterol		2.5	2.5	1.072	1.335		**
We should not eat beef so often		2.4	2.5	0.990	1.512		
(3) Taste							
Beef Tastes good		2.3	1.7	0.909	0.753	***	
Beef is the best tasting meat		2.7	2.9	1.315	1.590		
(4) Economy							
Beef is expensive		2.1	2.0	0.949	1.085		
We can't afford to eat beef every day		2.7	2.7	1.388	1.593		

Continued

Table 2.12. Comparative analysis of consumer's image of beef in Japan and the United States using the rating scale (RS) method (continued)

Item and attribute	Average ranking[a] 1 2 3 4 5 6	Average ranking Japan	Average ranking U.S.	Variance Japan	Variance U.S.	Significance level[b] Average ranking	Significance level[b] variance
(5) Role and importance in daily dietary life							
Beef is necessary for entertaining guests		2.7	3.3	0.930	1.561	***	***
Beef makes dinner elegant		2.4	2.7	0.887	1.175	***	*
Beef is indispensable for dinner		3.6	3.8	0.775	1.427		***
Beef should be eaten when dinning in a restaurant		4.1	3.7	0.850	1.078	***	
(6) Intentions							
We will eat more beef in the future		3.0	3.4	1.000	1.199	***	
We won't eat beef more often		2.9	3.1	0.910	1.360	**	**

Source: Study carried out by Tadashi Yoshida in 1983.

[a]One is most positive, 6 is least positive.

[b]The significance level is determined by the t-test with *** equaling the 0.01 level, ** equaling 0.05 level, and * equaling the 0.1 level level of probability.

beef higher and it appears they have more interest in increasing beef consumption than do U.S. consumers. However, the Japanese average marks are no more than 3.0 which denotes a neutral attitude.

Results of the SD technique and the RS method lead to the conclusion regarding the taste of beef that consumers of both countries respond positively. But, U.S. consumers, who eat the larger share of beef cooked by dry heat, feel more strongly about taste. The U.S. consumers view beef as an asset to human health but, at the same time, recognize its unhealthy characteristics. The perceived hazards and expense in consuming beef might be said to have caused the decrease in beef consumption. In contrast, Japanese consumers regard beef as a luxurious cuisine and appear to have a stronger intention of consuming more beef in the future than the U.S. consumers. Japanese only consume a relatively small amount of beef, but they apparently are beginning to perceive hazards of overeating beef and, accordingly, their intentions of consuming beef in the future are no more than moderate.

The Latent Factors Comprising Consumer's Beef Image

The latent factors comprising or affecting consumers' beef image of both countries was determined by applying factor analysis to the same data as in Table 2.12. The method used is to extract "common factors." The initial factor loading matrix is extracted by the principal factor method and then, rotating the factor axis by the normal varimax method, the final factor loading matrix derived. This matrix, provided in Table 2.13, has the factor loadings of 0.4 or more surrounded by rectangles, and those of 0.3 up to 0.4 by brackets.

As the table shows, 4 factors were extracted from the data of both the U.S. and Japanese consumers. Starting with U.S. consumers the first factor (I) has a correlation of 0.4 or more loading with 7 items: 4 items of "the role and importance of the daily diet," 2 items of "intention" and one item of "taste" (Beef is the best tasting meat") from Table 2.12. Putting these items together, we will name it "the factor of enriching daily diet." The 7 items with which the second factor (II) has 0.4 or more loading coincide exactly with those of "unhealthiness" from Table 2.12. Therefore, it can be called "the factor of unhealthiness."

The third factor (III) has a correlation of 0.4 or more loading with 4 items: 3 items of "healthiness" and one item of "taste" ("Beef tastes good.") from Table 2.12. It is called "the factor of healthiness." The 2 items with which the fourth factor (IV) has a correlation of 0.4 or more loading are the same as those of "economy" in Table 2.12 so it might be named

Table 2.13. Comparative analysis of consumer's image of beef in Japan and the United States using the factor analysis method

Item	United States				Japan				Communality	
	I	II	III	IV	I	II	III	IV	U.S.	Japan
Beef is indispensable for dinner	0.785	−0.039	−0.039	−0.091	0.681	0.087	−0.085	0.120	0.652	0.546
In the future we will eat more beef	0.730	−0.236	0.243	−0.054	0.412	0.139	0.058	−0.221	0.651	0.241
Beef is necessary for entertaining guests	0.727	0.156	−0.043	−0.084	0.534	0.103	−0.232	0.090	0.561	0.411
Beef should be eaten when dining in a restaurant	0.692	0.129	0.075	−0.068	0.032	0.153	0.041	0.420	0.506	0.256
We want to eat beef more often	0.610	−0.292	0.371	0.061	0.585	−0.106	0.424	−0.067	0.598	0.553
Beef makes dinner elegant	0.576	0.038	0.427	−0.029	0.690	0.054	−0.034	−0.060	0.517	0.484
Beef is the best tasting meat	0.519	−0.227	[0.314]	0.208	0.679	0.041	0.098	−0.021	0.462	0.475
Beef is rich in fat	−0.076	0.672	0.044	0.062	0.177	0.511	−0.168	−0.044	0.464	0.409
Beef makes people fat	0.010	0.666	−0.271	0.213	−0.054	0.545	0.057	0.054	0.563	0.306
We should not eat beef so often	−0.300	0.654	−0.202	0.262	−0.119	0.562	0.119	0.025	0.627	0.384
Consumption of beef causes high blood pressure	0.158	0.645	−0.169	0.119	0.061	0.601	−0.146	0.080	0.484	0.408
Beef tends to cause heart attacks	0.079	0.610	−0.203	0.192	0.107	0.459	0.074	[0.367]	0.457	0.375
Beef is full of cholesterol	−0.096	0.564	0.091	0.202	0.103	0.622	−0.130	−0.089	0.377	0.421
We should eat leaner beef	0.030	0.468	0.155	−0.104	0.164	0.237	0.008	[−0.344]	0.255	0.256
Beef is rich in nutrition	0.131	−0.021	0.728	−0.037	[0.396]	0.005	−0.089	−0.517	0.548	0.583
Beef tastes good	0.122	−0.134	0.625	−0.057	0.542	−0.094	0.162	−0.035	0.427	0.339
Beef is rich in vitamins	0.318	0.123	0.523	−0.032	[0.374]	−0.011	0.098	−0.160	0.391	0.176
Beef is good for health	0.257	−0.383	0.440	0.095	0.613	−0.069	−0.114	[−0.354]	0.415	0.560
We can't afford to eat beef every day	−0.044	0.177	0.011	0.733	0.123	−0.139	0.609	0.095	0.571	0.418
Beef is expensive	−0.093	0.345	−0.090	0.641	−0.033	0.035	0.561	0.029	0.547	0.324
Eigen value	3.477	3.217	2.122	1.255	3.270	2.000	1.088	0.959	—	—

Source: "Comparative Survey." Application of the factor analysis method. The initial factor loading matrix was extracted by the principal factor method and, by rotating the factor axis with the application of the varimax method, the final factor loading matrix was derived. The factor loadings with 0.4 or more are surrounded by rectangles, and those of 0.3 by brackets.

"the factor of economy." Thus, the latent structure of consumers' beef image might be said to be composed of 4 factors which are relatively distinct and evident.

The latent structure of the Japanese consumers' beef image is rather subtle compared with Americans. The first factor has a correlation of 0.4 or more loading with 8 items, of 0.3 to 0.4 with 2 items. These 10 items are composed of 3 from "the role and importance of the daily diet" (except one, "Beef should be eaten when dining in a restaurant"), 2 of "intention" and 2 of "taste" from Table 2.12. This seems to show that in the deep psychology of Japanese consumers the factor of enriching daily diet and that of healthiness are not yet separated. We will call the first factor "the factor of enriching daily diet and of healthiness."

Table 2.13 shows Japanese consumer's second factor mostly coincides with the third factor of U.S. consumers, the factor of unhealthiness; their third factor is nearly equal with the fourth factor of U.S. consumers, the factor of economy. The fourth factor is too obscure to be named as it includes such items as "Beef is rich in nutrition" and "Beef is good for health." But, it seems to be an offshoot of the factor of healthiness although it has a high correlation with the items of unhealthiness such as "Beef tends to cause heart attacks."

To sum up, we conclude that there are no large differences between the latent structures of consumers' beef images in the U.S. and Japan. Of course, some small differences are found especially in the deep psychology of Japanese consumers where the factor of enriching daily diet and that of healthiness are not yet separated. But it is worth noticing that, in the latent structure of Japanese consumers, the factor of unhealthiness is found distinctly.

International Consumption Comparisons

Beef consumption per capita in Japan is about one-eighth that of the United States and about a fifth the average in the ten countries which make up the European Economic Community (EEC). Pork consumption, in contrast, is about one-third that of both the United States and the EEC (Table 2.14). However, while poultry consumption is about one-third that of the United States, it is nearly equal to the EEC average. Overall, Americans eat 3.3 times as much red meat and poultry as in Japan, while consumers in the EEC take about 2.5 times as much. An important aspect is that in 1970 the ratios were 6.4 for the United States and 3.9 for the EEC. These data reinforce the conclusion given in the previous sections that Japan has moved quickly toward western food consumption patterns.

Table 2.14. Per capita consumption of red meats and poultry in selected countries, 1970 and 1983

Country	Beef and veal[a]	Pork	Sheep and goat meat	Poultry	Total	Total, ratio to Japan
	- - - - - - - - - - Kg - - - - - - - - - -					
			1970			
Japan	3.0	7.3	2.1	4.9	17.3	1.0
United States	53.3	33.2	1.5	22.2	110.3	6.4
EC-10 average	24.3	26.8	5.7	10.2	67.0	3.9
Germany, Fed. Rep.	25.1	38.4	0.2	8.4	72.2	4.2
Netherlands	19.7	26.5	0.2	5.7	54.6	3.2
			1983			
Japan	5.8	13.9	1.4	11.5	32.6	1.0
United States	49.0	30.2	0.8	29.6	109.6	3.4
EC-10 average	24.0	34.8	3.4	14.5	76.7	2.4
Germany, Fed. Rep.	21.7	50.4	0.8	9.4	82.3	2.5
Netherlands	17.5	37.2	0.4	11.0	66.1	2.0

Source: USDA, Foreign Agriculture Circular, Livestock and Meat, FLM 7-78 and FL&P 1-85.

[a]Carcass weight basis.

The use of per capita consumption to compare Japan with western countries is somewhat misleading because of considerable differences in body size. The Japanese per capita caloric intake is about 2,500 calories. In contrast it is about 3,200 in the United States. Also, when protein intake is compared, Japan is shown to be much higher in animal product intake than might be realized. For example, while United Kingdom consumers obtain 60 percent of their protein from animal products, Japanese get about 51 percent from animal products (Table 2.15). Americans obtain 67 percent of their protein from animal sources.

SUMMARY AND CONCLUSIONS

The Japanese diet has become moderately westernized in a short period of time. But, now that economic growth has been reduced, and an "optimal" diet in terms of protein, fat, carbohydrates and calories achieved, there is no compelling economic or dietary reason for the westernization process to continue. Thus, even though Japanese per capita consumption of red meat and poultry is just about one-third that in the United States, and beef consumption about one-eighth that of Americans, the only real reason for continued shift in consumption patterns will be due to changes in tastes and preferences.

Table 2.15. Daily per capita protein consumption in Japan and selected countries, 1978-1980

Country	Vegetable	Animal	Total
	- - - - - -	Grams	- - - - - -
Japan	43.3	45.1	88.4
France	37.6	67.4	105.0
Germany, Fed. Rep.	33.2	57.2	90.4
United Kingdom	36.3	55.0	91.3
United States	34.7	72.0	106.7
	- - - - - -	Percent	- - - - -
Japan	49	51	100
France	36	64	100
Germany, Fed. Rep.	37	63	100
United Kingdom	40	60	100
United States	33	67	100

Source: FAO Production Yearbook, 1982.

There are several conclusions arrived at from a comparative study between the U.S. and Japan. First is the determination of the extremely important role played by cooking methods on the decision to prepare a beef versus a non-beef dish. Cooking by dry heat, the major practice in the United States, leads to greater perception of flavor and, as a result, greater demand for the product. Western style kitchens with ovens and ranges as well as barbecue equipment are ever increasingly becoming standard features of Japanese households and that trend can be expected to continue. The new facilities will provide some impetus to expand beef consumption, but the effect will be relatively small.

A second piece of information obtained from the study is the extent to which Japanese take a mix of meat products during a meal compared with American eating styles. One reason is that prices of poultry and pork products are low compared with beef. But this is not a major reason for, although pork was 48 percent the price of beef in Japan during 1983 and it was about 71 percent in the United States, poultry was 32 percent the price of beef in both countries. Thus, a more important reason is taste and preparation methods.

Review of historical trends in Japan, and comparison of Japanese dietary patterns, would lead to an initial reaction that per capita consumption of beef will continue to grow steadily in Japan. But, that analysis does not take into account the "Essen-Kultur" or cultural factors. Review of Japanese as a people leads us to believe that while per capita consumption of beef will continue to grow, the rate will be modest. Our projections are provided in Chapter 9 at which time a rather detailed demand side analysis is carried out.

A third finding from analysis of cooking methods and beef as a food item is that beef must be considered as an ever-more diversified nonhomogeneous product. The process of diversification from the traditional methods which focused on thin sliced beef will probably continue. There are at least three levels of quality as reflected by price, and the trend has been toward greater consumption of lower priced cuts. But, the major reason has not been price. Rather, it is because tastes and preferences have changed to incorporate what coincidentially happened to be lower priced cuts in the diet.

Recognition that beef must be analyzed as a composite of various qualities is important for it points directions which Japan will take in incorporating technologies in beef production. The major message is that while most foreigners (and even many knowledgeable Japanese) believe the greatest demand (and consumption) is for heavily marbled beef, in reality the greatest market share -- and one which will grow -- is for the lower price cuts and, even in the high priced cuts, for a lower degree of marbling.

NOTES

1. This explanation was on the menu of a Japanese restaurant in Denver, Colorado.

2. See Sunset Easy Basics for Good Cooking, pp. 44 ff., and Rombauer and Becker Joy of Cooking, pp. 145-154, 442-448.

3. Longworth, Beef in Japan provides a detailed and exact description of the modernization of the dietary pattern and its factors. See Chapter 1.

4. The survey was performed by T. Yoshida in Gainesville with assistance of J. Simpson, in Chicago, of Prof. Kokoris (of Northeastern-Illinois University) and in Kyoto, of Asso. Prof Shimizu (of Nara Women's University) and Assist. Prof. Niiyama (of Kyoto University). In processing the data, much is owed to Assist. Prof. Oda (of Osaka Pref. University). Asso. Prof. W.N. Thurman (of North Carolina State University) and Dr. Y. Arayama read and improved the questionnaire of the survey and a manuscript on which this chapter is based.

Part II

Cattle Production in Japan: Historical Perspective and Current Situation

3
Livestock Industry Structure

Japan's cattle subsector, along with other livestock production subsectors, has changed as radically and quickly as the general farm structure described in Chapter 1. Most dramatic of all has been the decline in number and size of farms with beef cattle, as they fell from 1.4 million in 1965 to 315 thousand in 1984, a 10.5 percent annual average rate of decline during 1965-1975, and a 4.4 percent annual decline since then (Table 3.1). Farms with swine dropped from 702 thousand in 1965 to 92 thousand in 1984, while farms with broilers declined from 20 thousand to 7 thousand over this same time period. That rate of decline was 8.4 percent annually during 1965 to 1975 and 6.5 percent annually since 1965.

THE LIVESTOCK INDUSTRY IN TRANSITION

Total inventory of livestock and poultry underwent as dramatic changes from 1965 to 1984 as did farm numbers--but in the opposite direction. Dairy cattle inventory increased 64 percent from 1.3 million head to 2.1 million head. Swine numbers nearly tripled from 4.0 million head to 10.4 million head. Broilers increased eightfold, from 18 million birds to 143 million birds. However, beef cattle only increased a modest amount, 36 percent, from 1.9 million head to 2.6 million head (Figure 3.1).

The changes which took place from 1965-1984 are dramatized by placing inventory numbers on a per farm basis. Dairy cattle numbers increased about six times, from 3.4 head per farm to 24.1 head, and beef cattle increased from 1.3 head to 8.2 head. Swine inventory increased enormously, from 5.7 head per farm to 114 head. Broilers shot up nearly twenty-fold from an average of 892 birds per farm to 19,512 birds.

Table 3.1. Livestock and broiler farms, and inventory in Japan, 1965-1984

Year	Cattle: Dairy[a]	Cattle: Beef[b]	Swine	Broilers
	Farms[c]			
1965	381,600	1,434,580	701,560	20,490
1970	307,600	901,600	445,500	17,630
1975	160,100	473,600	223,400	11,540
1980	115,000	364,000	141,000	NA
1981	106,000	352,800	126,700	8,309
1982	98,900	340,200	111,800	7,715
1983	92,600	328,400	100,500	7,445
1984	87,400	314,800	91,500	7,330
	Head			1,000 birds
1965	1,288,950	1,885,810	3,975,960	18,279
1970	1,804,000	1,789,000	6,335,000	53,742
1975	1,787,000	1,857,000	7,684,000	87,659
1980	2,091,000	2,157,000	9,998,000	NA
1981	2,104,000	2,281,000	10,065,000	131,252
1982	2,103,000	2,382,000	10,040,000	130,585
1983	2,098,000	2,492,000	10,273,000	134,612
1984	2,110,000	2,572,000	10,423,000	143,024
	Head per farm			Birds per farm
1965	3.38	1.31	5.67	892
1970	5.86	1.98	14.22	3,048
1975	11.16	3.92	34.40	7,596
1980	18.18	5.93	70.91	NA
1981	19.89	6.47	79.44	15,796
1982	21.26	7.00	89.80	16,926
1983	22.66	7.59	102.22	18,081
1984	24.14	8.17	113.91	19,512

Source: Adapted from MAFF. Statistical Yearbook of Ministry of Agriculture, Forestry and Fisheries. Tokyo, various issues.

[a]Female.

[b]Includes dairy cattle for fattening.

[c]As of February 1.

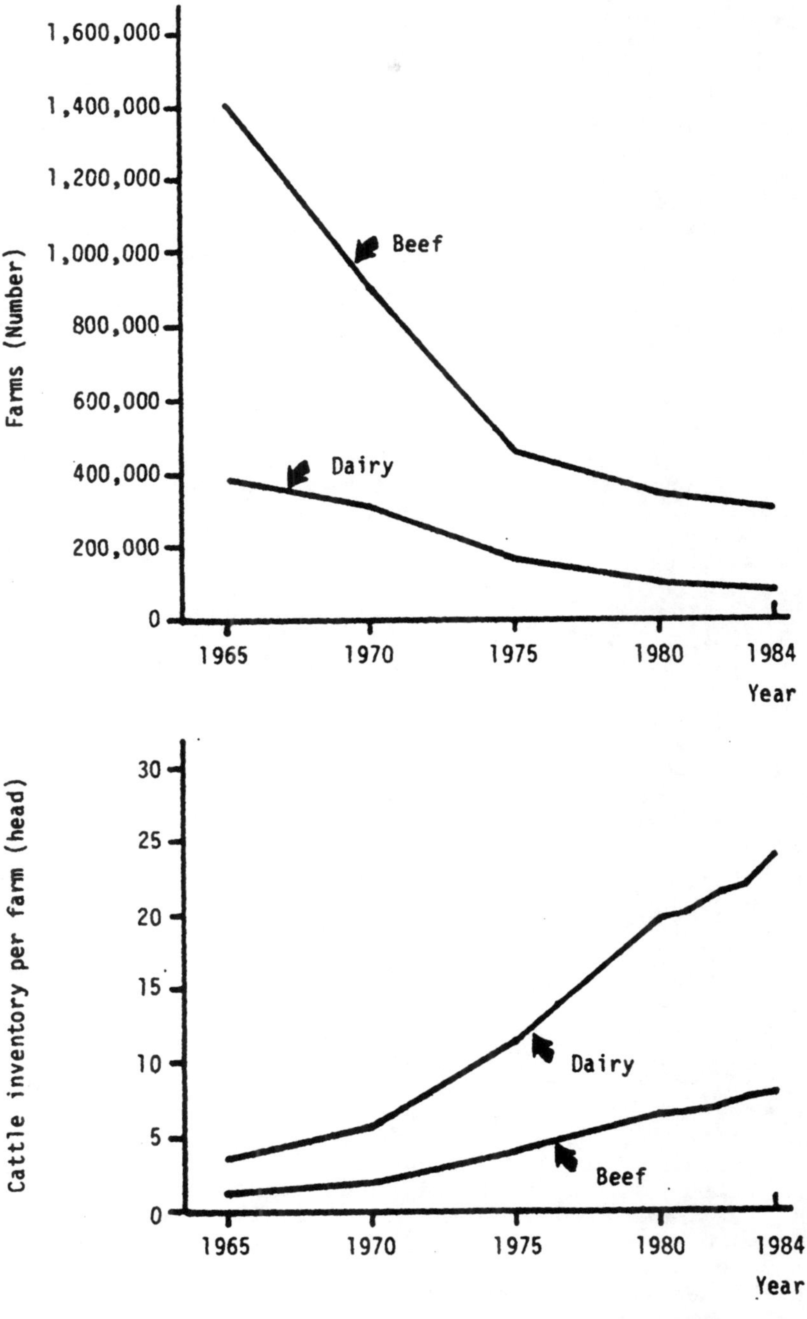

Figure 3.1 Number of dairy and beef farms and cattle inventory per farm, Japan 1965-1984

The transition in Japan from being characterized by small, subsistence, integrated crop/livestock farms in the early 1960's, to the much more specialized operations found today, is continuing. In just 8 years, from 1975-1984, the number of farms with beef cattle declined by one-fourth, from 474 thousand to 315 thousand. The net result is that beef cattle inventory per farm doubled in this 8 year period, from 3.92 to 8.17 head (Figure 3.1).

Distribution of cattle on farms has changed in concert with average inventory per farm increasing rapidly. In 1975 almost exactly two-thirds of the beef cattle were on farms with 1-2 head; by 1984 those very small units had fallen to less than half of the total (Table 3.2). In 1975 only 0.7 percent of the farms had 50 head or more; by 1984, 2.7 percent of all farms had more than 50 head including 1.1 percent of the farms which had 100 or more head. In absolute terms, the number of farms in these largest categories more than doubled, from 3,545 in 1975 to 8,460 in 1984. This is an average increase of ten percent annually!

Analysis of beef cattle inventory distribution by farm size provides even more dramatic results than the above information on the number of farms in each category. Also, it sheds new light on changes that are taking place in size of Japan's cattle operations. For example, in 1975 about 26 percent of all beef cattle were on farms with 1-2 head or cattle (Table 3.3). But, by 1984 only 9.5 percent were on these smallest operations. Furthermore, inventory of beef cattle on farms with 4 head or less fell from 42 percent of all beef cattle in 1975 to just 21.3 percent in 1984.

The growth in inventory on farms with 50 head or more of beef cattle has been as impressive as the decline in small operations. In 1975, 14.2 percent of all beef cattle were found on farms with 50 head or more. By 1984 the total had grown to 31.9 percent, or more than doubling. In effect, Japan is rapidly moving toward a more western hemisphere commercial type beef cattle production structure. Care must be taken in interpretation of the data, as dairy steer and heifer fattening for slaughter are included in statistics called "beef." Thus, a substantial portion of the larger units have dairy rather than beef breed cattle.

Geographically, the heaviest concentration of farms raising beef cattle is found in the Southern part of Kyushu. Two other fairly heavy areas of concentration are Sanin and Tohoku. Total cattle inventory, on the other hand, is fairly well distributed throughout Japan with the exception of Southeastern Kyushu which has the heaviest concentration (Figure 3.2). The relatively large size of farms in Hokkaido is reflected by the small concentration of them in that prefecture, but inventory being on par with the rest of Japan.

Table 3.2. Number of farm households raising beef cattle by size, Japan, 1965–1984

Year[a]	Total farms	Number of farms of raising beef cattle by size of herd: 1–2	3–4	5–9	10–19	20–29	30–49	50–99	100 and more
					Farms[b]				
1975	473,600	315,300	86,080	44,360	14,760	5,800	3,759	3,545	
1976	449,600	282,900	90,510	46,024	16,413	5,936	3,850	2,722	1,242
1977	424,200	257,800	86,254	47,915	17,294	6,011	4,335	3,152	1,502
1978	401,600	240,700	79,700	48,500	16,800	5,930	4,430	3,630	1,900
1979	380,800	219,900	77,500	48,000	17,800	6,480	4,950	3,860	2,250
1980					NA				
1981	352,800	185,200	78,800	51,800	18,300	6,560	4,960	4,500	2,800
1982	340,200	167,900	79,800	53,800	19,200	6,800	5,020	4,600	3,000
1983	328,400	157,600	77,800	53,200	19,900	6,970	5,150	4,620	3,260
1984	314,800	143,800	76,500	53,300	20,200	7,070	5,500	4,930	3,530
					Percent				
1975	100	66.6	18.2	9.4	3.1	1.2	0.8	0.7	
1976	100	62.9	20.1	10.2	3.7	1.3	0.9	0.6	0.3
1977	100	60.8	20.3	11.3	4.1	1.4	1.0	0.7	0.4
1978	100	59.9	19.8	12.1	4.2	1.5	1.1	0.9	0.5
1979	100	57.7	20.3	12.6	4.7	1.7	1.3	1.0	0.6
1980					NA				
1981	100	52.5	22.3	14.7	5.2	1.9	1.4	1.4	0.8
1982	100	49.4	23.5	15.8	5.6	2.0	1.5	1.4	0.8
1983	100	48.0	23.7	16.2	6.0	2.1	1.6	1.4	1.0
1984	100	45.7	24.3	16.9	6.4	2.2	1.8	1.6	1.1

Source: Derived from MAFF. Statistical Yearbook of Ministry of Agriculture, Forestry and Fisheries. Tokyo, 1984.

[a]As of February 1.

[b]Includes farms fattening dairy cattle for beef.

Table 3.3. Estimated total number of beef cattle by herd size per farm, Japan, 1975–1984

Year	Total cattle being raised[a]	Average number of cattle per farm	Estimated number of beef cattle being raised by herd size per farm: 1–2	3–4	5–9	10–19	20–29	30–49	50–99	100 and more
						Head				
1975	1,853,454	3.9	472,950	301,280	310,520	214,020	142,100	148,500	264,084	
1976	1,925,788	4.3	424,350	316,785	322,168	237,989	145,432	152,075	202,789	124,200
1977	1,978,284	4.7	386,700	301,889	335,405	250,763	147,270	171,233	234,823	150,200
1978	2,003,805	5.0	361,050	278,950	339,500	243,600	145,285	174,985	270,435	190,000
1979	2,062,055	5.4	329,850	271,250	336,000	258,100	158,760	195,525	287,570	225,000
1980						NA				
1981	2,153,440	6.1	277,800	275,800	362,600	265,350	160,720	195,920	335,250	280,000
1982	2,193,740	6.4	251,850	279,300	376,600	278,400	166,600	198,290	342,700	300,000
1983	2,214,030	6.7	236,400	272,300	372,400	288,550	170,765	203,425	344,190	326,000
1984	2,260,200	7.2	215,700	267,750	373,100	292,900	173,215	217,250	367,285	353,000
						Percent				
1975	100		25.5	16.3	16.8	11.5	7.7	8.0	14.2	
1976	100		22.0	16.4	16.7	12.4	7.6	7.9	10.5	6.4
1977	100		19.5	15.3	17.0	12.7	7.4	8.7	11.9	7.6
1978	100		18.0	13.9	16.9	12.2	7.3	8.7	13.5	9.5
1979	100		16.0	13.2	16.3	12.5	7.7	9.5	13.9	10.9
1980	100						NA			
1981	100		12.9	12.8	16.8	12.3	7.5	9.1	15.6	13.0
1982	100		11.5	12.7	17.2	12.7	7.6	9.0	15.6	13.7
1983	100		10.7	12.3	16.8	13.0	7.7	9.2	15.5	14.7
1984	100		9.5	11.8	16.5	13.0	7.7	9.6	16.3	15.6

Source: Average inventory per farm calculated by dividing the total inventory in Column 1 by total farms in Column 1 of Table 3.2.

[a]Estimated inventory by size classification obtained by multiplying total number of farms by mean inventory for each category in Table 3.2. For example, in 1975, there were 315,300 farms in the category of 1–2 head per farm. Multiplying 315,300 times 1.5 (the mean of a 1–2 head category) yields 472,950 head. The totals may not add due to rounding.

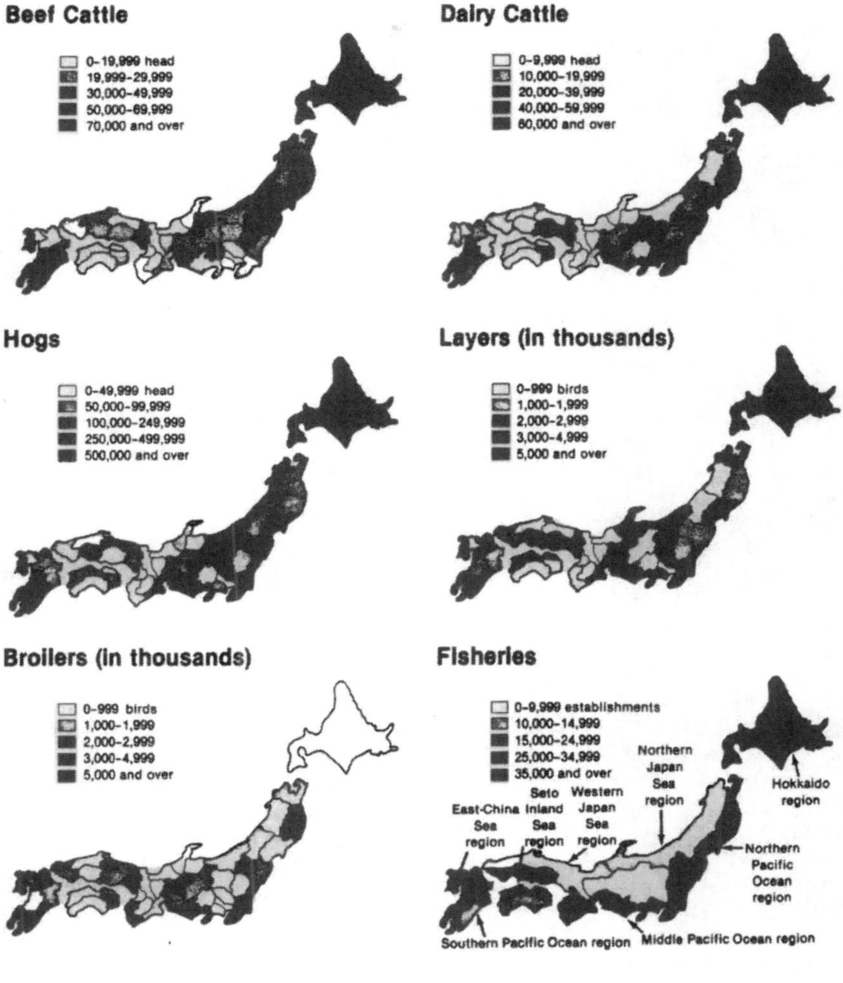

Figure 3.2 Distribution of livestock and fisheries by prefecture, Japan, 1983

Source: Coyle, William T. Japan's Feed Livestock Economy: Prospects for the 1980's. Washington, D.C.: USDA FAER 177, February, 1983.

About 70 percent of Japan's beef and veal is derived from dairy cattle, making this source of beef quite important in analysis of beef production potential. In contrast, only about 14 percent of beef and veal consumed in the United States is derived from dairy cattle. Beef cattle inventory, by class of cattle, is given in Table 3.4.

The dairy industry of Japan, just like beef production as an industry per se, is relatively new since milk and dairy products did not begin to enter into general diets until the 1920's and not to any great extent until after World War II. Improved breeds of milk cattle were introduced into Japan during the Tokagawa period (1603-1867), probably by the Dutch. But, the early period milk was mainly used by evaporating it and ingesting the concentrate as a drug for treating tuberculosis and syphilis.

In the early 1900's dairy herds were primarily maintained in suburban areas, with replacements often coming from the more remote areas. Today, Japan's dairy industry is quite well developed with yields approaching those in major milk producing countries. For example, in 1983 the average annual production per cow according to FAO data was 4,790 kg in Japan compared with 5,637 kg in the United States, 5,333 kg in the Netherlands, and 6,955 is Israel.

The greatest changes on dairy farms have taken place in the past few years for, while in 1975 the total number of farms had fallen to 160 thousand with 11.2 head per farm (from 410,000 farms in 1960) only 9 years later in 1984 the number of farms had decreased by 45 percent to just 87 thousand units, but with an average of 24.1 head per farm (Table 3.5). Total dairy cattle numbers in the meantime have been growing rapidly and continuously, from 1.3 million head in 1965 to 2.1 million head in 1984. Japan is about 83 percent self-sufficient in dairy products with imports mainly being cheese and skimmed milk powder for feed.

Dairy production location within the country shifted rapidly after World War II from the traditional suburban farms to more distant rural areas where grazing land is more abundant. This was mainly due to urbanization and developments in the transportation infrastructure which permitted milk to be hauled economically over much greater distances. In 1965 only a small amount of milk was produced in the extreme northern island of Hokkaido; now that area has the largest concentration of dairy cattle and accounts for about one-third of all production (Figure 3.2). Hokkaido farmers could easily expand to a much larger production if a market were available. Per capita consumption of dairy products has increased continuously, from 22.2 kg in 1960 to 62.1 kg in 1980 and 67.1 kg in 1984 (Appendix 3.1). Despite this, one of the more serious policy

Table 3.4. Beef cattle inventory in Japan, 1965–1984

Year[a]	Total	Female		Male		Dairy cattle for beef
		Under 2 years	2 years and older	Under 2 years	2 years and older	
1965	1,434,580	394,750	919,090	571,970		NA
1970	1,789,000	370,750	794,400	624,250		NA
1971	1,759,000	374,800	714,200	367,500	116,300	186,300
1975	1,857,000	340,100	609,300	331,500	100,700	475,500
1980	NA	NA	NA	NA	NA	NA
1981	2,281,000	353,300	632,600	378,500	113,300	803,300
1982	2,382,000	375,100	642,700	393,500	118,100	852,600
1983	2,492,000	391,800	671,200	418,500	125,000	885,800
1984	2,572,000	408,300	678,000	444,700	126,800	913,900

Source: MAFF. <u>Statistical Yearbook of Ministry of Agriculture, Forestry and Fisheries</u>. Tokyo, March, 1985.

[a]As of February 1.

Table 3.5. Dairy cattle inventory, average herd size and milk production in Japan, 1960–1984

Year	Dairy farms	Total dairy cattle	Number of cows that have calved and are in lactation	Average herd size	Raw milk production			Annual milk production per cow
					Fresh	Processed	Total	
	- No -	- - 1,000 Head - -		- Head -	- - - - - 1,000 MT - - - - -			- Kg -
1960	410,420	824	NA	2.0	987	742	1,887	NA
1965	381,600	1,289	634	3.4	1,771	1,259	3,221	5,100
1970	307,600	1,804	885	5.9	2,623	1,962	4,761	5,400
1975	160,100	1,787	910	11.2	3,130	1,713	4,961	5,500
1976	147,100	1,811	928	12.3	3,314	1,830	5,262	5,700
1977	136,500	1,888	968	13.8	3,548	2,067	5,735	5,900
1978	129,400	1,979	1,013	15.3	3,682	2,310	6,117	6,000
1979	123,300	2,067	1,072	16.8	3,857	2,468	6,463	6,000
1980	NA	NA	NA	NA	3,988	2,334	6,504	NA
1981	106,000	2,104	1,075	19.8	4,121	2,303	6,610	6,100
1982	98,900	2,103	1,082	21.3	4,215	2,385	6,747	6,200
1983	92,600	2,098	1,096	22.7	4,272	2,660	7,042	6,400
1984	87,400	2,110	1,101	24.1	4,323	2,709	7,141	6,500

Source: MAFF. Statistics of Milk and Milk Products. Tokyo, various issues.

matters in Japan today is reestablishment of a fluid milk program to adjust regional milk production and thus avoid oversupply.

FEEDER CATTLE PRODUCTION INDUSTRY STRUCTURE

Feeder cattle, i.e. calves which will go into feedlots for fattening to slaughter weight, are essentially a by-product of the dairy industry, but the primary product of the beef industry. Data are not available to determine the number of farms specializing in beef breed feeder cattle production or in their size classification. However, it appears that farms falling in the smaller herd classifications (Table 3.3) tend to be cow/calf operations while the larger herd per farm classifications are likely to be confinement cattle feeders. Thus, a typical cow/calf operation would have about 3 cows, while a feeder would have 50 or more head. Wagyu calves are weaned at fairly heavy weights, about 285 kg.

There are some integrated beef breed cow/calf and fattening operations but, typically, calves are sold after weaning at about 8 months of age. Virtually all cow/calf operations have Wagyu breed cattle. In general the calves, which are halter-broken and quite docile, will be transported to an auction market where the producer will clean, brush and groom them, much like show cattle are prepared at a United States fair. They are then sold individually in an auction ring. The pedigree of each animal is available for inspection by potential buyers and appears to influence price equally as much as phenotypic characteristics.

Dairy calves are procured and raised for placement in feedlots under three systems. One is the dairy itself raising the calf on milk replacer to 7-8 months of age (250-270 kg) at which time it is sold to a feedlot. There are virtually no dairy operators who are also integrated into cattle feeding. The second system, and probably the most common, is sale of newborn (1-7 days of age) calves directly to feedlot operators who raise them in calf hutches on milk replacer. The third system involves a specialized operator who buys calves from dairies at 7-8 days of age, grows them out, and sells to feedlots at 7-8 months of age. In all systems death loss is relatively high (usually 10-20 percent) because of poor management by dairymen, mainly from not providing proper environmental conditions and/or feeding adequate amounts of colostrum, i.e. new milk from the mother.

FEEDING CATTLE TO SLAUGHTER WEIGHTS

The role of beef breeds, about 96 percent of which are Wagyu (Appendix 3.2), has changed remarkably since World War II and it appears their role will continue to change. Wagyu were primarily used for draft power through the 1940's and into the 1950's, at which time mechanization replaced them (horses were the main source of draft power in some areas such as Hokkaido rather than cattle). Because males were preferred for draft, fattening for beef was heavily oriented to females, and primarily virgin heifers. In fact, the famous Kobe beef is derived from that practice.

During the 1960's and into the 1970's the beef cattle industry developed quickly, with the new primary purpose being beef output. As a result, specialized cow/calf operations and feedlots developed so that, within a decade, a major transition had been made. As Japanese per capita incomes improved and the country industrialized (with resultant expanded expense accounts), the demand for highly marbled beef grew at a rapid rate. The net result was large price differentials between beef grades. As might be expected, feeding strategies by beef breed producers became heavily oriented toward reaching the highest grade possible. That led to expanded time on feed and deliberate slow down of average daily gain as a means to improve marbling.

Now, in the mid-1980's, recognition is developing among many producers and others allied with the industry such as Ministry of Agriculture officials and researchers that (1) the market for beef is becoming much more differentiated, and (2) that the most efficient means of beef production (from both a cost and resource use standpoint) is to group cattle according to their genetic potential and feed them accordingly. That means some cattle would be targeted to the two highest possible grades, Supreme and Superior (Tokusen and Goku-Jyo) while the majority would be fed much shorter periods with efforts concentrated on achieving the highest possible average daily gain. In effect, whereas the feeding strategies are now relatively homogeneous (even though size of operation varies greatly), the next steps will be much more oriented to changing market and competitive conditions.

Dairy steer fattening emerged as an industry in the late 1960's and became well established in the early 1970's. One way to track that development is by analysis of dairy calf slaughter statistics. In 1969-1970 there were about 310-320 thousand dairy calves slaughtered. Numbers declined rapidly to 1979 when 41 thousand head were slaughtered. The feedlot industry can be considered "mature" as of that time since calf slaughter has remained at about 41 thousand head since then. There are now about 460 thousand finished dairy steers slaughtered annually along with about 480 thousand cows and heifers (Appendix 3.3).

Wagyu steer slaughter ranged between 143 thousand head and 251 thousand head from 1965-1983 (Appendix 3.4). It reached a 20 year record in 1984 of 269,822 head. Calf slaughter has been relatively constant since 1971 indicating (1) the beef breed feeding industry has been fairly well-established since that time, and (2) the long-term importance of beef breeds for fattening.

Comparison of data in Appendix 3.3 and 3.4 shows that while approximately equal numbers were slaughtered in the early 1970's, by the early 1980's there were about twice as many dairy cattle slaughtered as beef cattle. Twenty years ago, in 1965, there were 2 beef cattle slaughtered for each dairy breed animal.

The data on slaughter mix, along with information about changes in size of farms, and livestock inventory per farm, demonstrate the rapidity with which adjustments are being made in Japan's beef feeding industry. Cattle feeding on a commercial scale is a relatively new phenomenon in Japan, only having developed in the late 1960's, with solid growth not taking place until the 1970's. Data are not available on the number or distribution of Wagyu cattle feeders, but observation and interviews indicate that the greatest number of cattle are finished in lots with 50 head or more even though the majority of farms only have a few head. In large part, the growth of this size category (Table 3.3) reflects progress in development of the feeding industry. The 14 percent annual growth rate in Japan of large commercial operations (as measured by growth in farms with 100 head or more) is comparable to the development of cattle feeding on the High Plains of the United States in the early 1960's.

SUMMARY AND CONCLUSIONS

Japan's livestock industry is changing dramatically, both in practices and structure. Broiler and swine units are now very large and among the world's finest. Changes will continue to take place, but these industries can be considered as virtually mature ones. Cattle production and marketing have changed more in the early 1980's than in any previous equivalent length period. In fact, the early to mid 1980's could be called the "age of diversification" in the Japanese beef industry. It is only by recognizing and understanding the great amount of change taking place that the industry, on both the demand and supply sides, can be understood. Japanese cattle units are still relatively small on the average, but many large scale dairy farms and feedlots have emerged. The one industry still characterized by small units is Wagyu cow/calf production. Even there consolidation is rapidly taking place, as the total number of farms with beef cattle dropped from 474 thousand in 1975 to 315 thousand in 1984. Most of that change was in units of 1-2 head where farm numbers dropped from 67 percent of the total in 1975 to 48 percent in 1984.

Change in industry structure is a painful experience to both those operators exiting as well as the ones struggling to become more competitive. In general, small operations tend to be owned by older people and thus, to a great extent, these farms will just cease to exist as the owners retire. In most cases the land will be consolidated into larger holdings. On the attrition side younger and middle aged farm families are often faced with heavy debt loads. They can often see that expansion is necessary to survival, but cash flow problems prevent rapid change. It is these and other types of micro and macro level policy problems which face Japan today.

4
Cattle Feeds and Feedstuffs

The two major expense items in beef production are the animal and feed. Because only 13 percent of Japan's land area is arable, expansion of beef production, along with pork, poultry and dairy products, has required continual growth in imports of feedstuffs. As of 1960 about 10.4 million tons of nonpasture feeds were utilized, and Japan was 63 percent self-sufficient in them (Table 4.1). But, by 1975 total utilization had doubled, and the ratio was down to 34 percent. The downward trend in self-sufficiency about leveled off by 1977 with the ratio at about 30 percent and imported nonpasture feed material at 15 million tons. An even more dramatic change has taken place in feed grain use.

FORMULA FEEDS

Formula feeds, i.e. ones with 2 or more ingredients now constitute about 60 percent of national feed supply on a total digestible nutrient (TDN) basis, and about 80 percent of total concentrates fed. The greatest increase in use of formula feeds per livestock unit (including poultry) has taken place in beef cattle, where consumption increased from 15 kg in the 1960-1965 period, to 1,223 kg by 1980 (Table 4.2). Beef cattle formula feed use grew at 13.2 percent annually during 1975-1980 compared with 7.8 percent for all formula feed. Much of this additional feed is for finishing both beef and dairy cattle to slaughter weights. Corn is the major ingredient in formula feeds for all livestock (nearly 50 percent). It accounts for about one-third of beef cattle formula feeds. Sorghum is second for both beef cattle and all livestock (Appendixes 4.1 and 4.2).

Simple feeds, farm mixed and food-processing by-products have continuously been replaced by formula

Table 4.1. Nonpasture feed supply in Japan, 1960-1984

Calendar Year	Total utilization (A)	Roughage (B)	Total concentrate feed (C)	Domestically produced material (D)	Imported material processed (E)	Imported material (F)	Self-sufficiency $(\frac{B+D}{A}) \times 100$
	- - - - - - - - - - - - - - - 1,000 Metric Tons	- - - - - - - - - - - - -					- Percent -
1960	10,423	4,877	5,671	1,683	2,090	1,898	63
1965	13,359	4,519	8,840	2,771	1,136	4,932	55
1970	18,394	4,656	13,739	2,297	2,176	9,266	38
1975	19,867	4,793	15,074	2,060	2,639	10,375	34
1976	21,402	4,815	16,587	1,944	2,690	11,452	32
1977	22,782	4,879	17,903	1,844	2,805	13,255	30
1978	24,114	5,181	18,933	1,792	3,102	14,039	29
1979	25,529	5,175	20,354	1,888	3,181	15,285	28
1980	25,107	5,118	19,989	1,965	3,038	14,986	28
1981	24,899	5,168	19,731	2,283	3,180	14,268	30
1982	25,491	5,441	20,050	2,694	3,217	14,139	32
1983[a]	26,069	5,206	20,863	2,538	3,313	15,012	30
1984[b]	26,524	5,593	20,931	2,128	3,321	15,482	29

Source: Format from Coyle, W.T. Japan's Feed-Livestock Economy. FAER No. 177, USDA, Washington, D.C., 1983, and based on data from the Japan Feed Association, Yearly Report on Cencentrate Feed. Tokyo, various issues.

[a]Tentative.

[b]Estimated.

Table 4.2. Formula feed consumption per livestock unit in Japan, 1960–1980

Type of livestock	Year				
	1960–65	1965–70	1970–75	1975–80[a]	1980
	- - - - - - - - - - - - Kilograms - - - - - - - - - - - -				
Beef cattle	15	230	789	1,088	1,223
Swine	282	500	627	677	640
Broilers	NA	NA	5.2	5.4	5.4
Layers	45.9	45.0	43.9	45.7	46.0
Dairy cattle	530	823	1,021	1,101	1,088
	- - - - - - - Annual Percentage Change - - - - - - -				
Beef cattle	NA	62.6	12.0	13.2	
Swine	NA	17.3	2.9	9.2	
Broilers	NA	27.0	9.0	9.3	
Layers	NA	7.4	-1.2	2.6	
Dairy cattle	NA	16.7	1.0	6.4	
Total formula feed	23.2	13.0	2.3	7.8	

Source: Coyle, W.T. Japan's Feed-Livestock Economy. FAER No. 177, USDA, Washington, D.C. 1983, and based on data from Japan Feed Association Shiryo Geppo (Feed Monthly) and MAFF, Meat Statistics of Japan. Tokyo, various issues.

[a]Annual percentage change is for 1975–1979.

feeds on a percentage basis. However, on an absolute basis, use of simple feeds has continued to increase as producers, especially cattle feeders, have sought to mix their own rations. Barley, corn and wheat bran are the major ingredients. All animal feeds have been imported duty-free since 1927. Importing feed mills, usually located near ports, had to be licensed. Steam flaked corn has become more important since 1976 due to a Ministry of Agriculture and Forestry (MAFF) regulation at that time which permitted it to be imported duty-free provided it was denatured (mainly with fish meal), a composition call Nishukon. (It is denatured to prevent it from being diverted to human consumption.) Animal feedstuffs, provided they are imported by bonded mills, have an automatic approval system, no quota, and no tariff.

FEED GRAIN PRODUCTION, IMPORTS AND POLICIES

The Japanese government has encouraged expanded domestic production of feed grains and roughages. One means has been the use of subsidies under all three of the rice diversion programs during the 1970's in which farmers received large payments for planting feed grains on rice land. Considerable research has been undertaken on feed grain production, and it appears likely that yields will increase substantially in the 1980's even though total tonnage of domestically produced materials (grain and by-products, fish meal, and nonfat dry milk) are actually at a lower total level than in the mid-1960's (Table 4.1).

Imports of coarse grains grew from 1.4 million tons in 1960 to 20.7 million tons in 1984 due to rapid expansion in feeding livestock (Table 4.3). Care must be taken, as shown in Appendix 4.3 to separate human from livestock use. United States share, which only constituted about 16 percent of Japan's imports in 1960, shot up to 84 percent by 1980. It has fluctuated between 72 and 78 percent since then (Appendix 4.4). Oilseed imports have also continued to increase, reaching 6.4 million tons in 1983 although they fell to 6.1 million tons in 1984. Soybeans account for about three-fourths of oilseed imports (Appendix 4.5).

A major concern of government is to use livestock feeds as a device to help control livestock price fluctuations. A number of methods are used such as state trading, stockpiling, supply/purchase agreements, supply source diversification and a price stabilization fund. Overall, it appears that the governments' influence through state trading is diminishing. Also, stockpiles are small in relation to annual feed consumption. A feed price stabilization fund established

Table 4.3. Imports of coarse grains by Japan, 1960–1984[a]

Calendar year	Corn	Sorghum	Barley	Oats	Rye	Millet	Total
			1,000 Metric Tons				
1960	1,354	45	—	2	3	—	1,404
1965	3,434	1,431	635	13	46	69	5,628
1970	6,018	3,789	768	135	73	43	10,826
1975	7,470	3,794	1,598	141	54	49	13,106
1980	12,830	4,219	1,416	117	18	47	18,707
1981	13,590	3,301	1,568	114	18	48	18,639
1982	13,571	3,428	1,330	95	20	53	18,497
1983	14,701	2,957	1,477	74	133	48	19,270
1984	14,170	4,478	1,567	126	342	36	20,719
			Percent[b]				
1960	96	03	—	—	—	—	100
1965	61	26	11	00	01	01	100
1970	56	35	07	01	01	00	100
1975	57	29	12	01	01	00	100
1980	74	23	08	01	00	00	100
1981	73	18	08	01	00	00	100
1982	73	19	07	01	00	00	100
1983	76	15	08	00	01	00	100
1984	68	22	08	01	02	00	100

Source: Derived from Ministry of Finance. Japan Exports and Imports. Tokyo, annual issues and MAFF. Abstract of Statistics on Agricultural, Forestry and Fisheries, 1984. Tokyo, March, 1985.

[a]Includes feeds for both animal and human consumption. See Appendix 4.4.

[b]May not add to 100 due to rounding.

in 1975 is used to reduce short-term fluctuations in feed prices paid by livestock producers.

Supply-purchase agreements between Japan and exporting countries are an important part of policy. The most important agreement was the Butz-Abe understanding resulting from a 1975 meeting. It expired in 1978 and no similar arrangement has replaced it despite numerous meetings and a strong desire for one of the part of the United States.

MAFF AND USDA FEEDGRAIN USE PROJECTIONS

The demand for feedstuffs is based on consumption of products produced by them. The projections of animal feed needs now presented are from the United States Department of Agriculture (USDA), and only cover 1985 and 1990. The first step to determine animal feed needs is to calculate use of feed for livestock products. These projections, based on ones derived by _Shiryo Geppo_ (_Feed Monthly_) and _The Meat Statistics of Japan_, indicate that while 8.1 kg of concentrate was used per kilo of beef and veal in 1980, the ratio will increase to 11.7 kg in 1990 (Appendix 4.6). These ratios are _not_ feed conversions, but rather are indicators of total feed use as they are obtained by dividing total feed by total animals. Concentrate use will grow due to expanded use of feedlots and intensified production methods. Pork, which is already produced mainly by concentrates, will have its per unit feed use only grow slightly (4.34 kg to 4.59 kg per kg of product). The per unit feed use for chicken, eggs and milk will decline as both farm size and efficiency increase.

USDA projects feed grain demand will grow between 4.5 and 5.7 percent annually from 1980-1990. In contrast, MAFF projects it at 3.3 percent (Table 4.4). USDA projections forecast that whereas domestic production of coarse grains was about 400 thousand metric tons in 1980, it will reach 577 thousand tons in 1990. Total consumption of coarse grains in 1980 (of which major livestock feeds constitutes 82 percent) was 19.2 million tons, while imports were 18.9 million tons or 98 percent of total consumption (Appendix 4.7). The high USDA projections reckon consumption will be 32.8 million tons by 1990, which means imports would be 32.2 million tons if they constitute 98.2 percent of the total. The above data represents a 5.5 percent annual rate of growth.

Protein meal rates are expected by USDA to grow between 3.8 and 5.0 percent annually during the 1980's. In contrast, the MAFF projection is 2.2 percent. Given this trend, and taking into account growth in livestock being fed, USDA projects that protein meal demand for

Table 4.4. Japan: Projections of feed grain demand in Japan, 1990

Item	Livestock production in 1980	Grain demand	MAFF projections[1] Livestock production	MAFF projections[1] Grain demand	ERS projection I Low Livestock production	ERS projection I Low Grain demand	ERS projection II High Livestock production	ERS projection II High Grain demand
	1,000 Metric Tons							
Beef and veal	418	2,391	630	52,298	529	4,449	596	5,012
Pork	1,476	4,679	1,940	6,674	2,498	8,593	2,799	9,629
Chicken	1,145	2,279	1,460	2,935	1,868	3,755	2,112	4,145
Eggs	1,999	5,037	2,220	5,017	2,398	5,419	2,695	6,091
Dairy products	6,502	1,365	8,420	1,937	9,298	2,139	10,485	2,412
Total grain demand	--	15,751	--	21,861	--	24,355	--	27,389
Average annual increment, 1980-1990	--	--	--	611	--	860	--	1,164
	Percent							
Average annual growth rate, 1980-1990	--	--	--	3.3	--	4.5	--	5.7

-- = Not applicable.

Source: Coyle, W.T. Japan's Feed Livestock Economy. FAER No. 177. USDA, Washington, D.C., 1983.

[1]Based on MAFF's The Long-Term Prospects for the Demand and Supply of Agricultural Products, released by the Government of Japan, November, 1980.

beef and veal production will increase from 234 thousand metric tons in 1980 to between 428 and 483 thousand tons in 1990, i.e. about doubling over the 10 year period (Appendix 4.8). Virtually all of this would come from expanded imports of both oilseeds and meal (Appendix 4.9).

SOME BASIC ECONOMICS

There are many reasons why a country or region would be concerned with maintaining or expanding what amounts to only a modest level of self-sufficiency in beef production. In Japan's case one major reason is domestic security. Also, having a reserve of feed grains as well as livestock on farms and in feedlots offers a greater amount of food in reserve than simply importing meat directly. The income and employment generating factors are important, but more critical is a natural inclination to protect an established industry, especially one as vociferous as farmer's groups in Japan. Another reason is preference for domestically produced products. In the case of Japan this is partially true because of a desire for fresh rather than frozen meat.

At first glance it would appear that Japan is severely disadvantaged by importing grain rather than beef. However, the ocean freight rates on a feed grain equivalent basis turn out to be similar. For example, in mid-1982 grain transportation cost was about $26 per ton (actual weight basis) from U.S. Gulf ports to Japan, while frozen meat in cartons was about $300 per ton. On a feed grain equivalent basis bone-in frozen meat transportation charges were about $33 per ton and boneless frozen meat $22 per ton (Longmire, 1984). Another study by Gorman and Mori (1985) demonstrates that longer term feeding in the United States to reach high prime as a substitute for the best quality Japanese beef market would be marginal at best. The point is that from a Japanese point of view costs on a meat equivalent basis are about similar. Thus, questions related to economics of feedgrains hinge on efficiency in use of grains, policy, production costs and other supply side related factors.

CROP AND FORAGE PRODUCTION TECHNOLOGIES

Japan has, as shown in the last several sections, increasingly depended on imported feedstuffs for livestock production. However, now that rice production has become a surplus problem, and sufficient advances have been made in production techniques of non-animal feedstuffs so the the country has achieved a comfortable self-sufficiency ratio in them, it is natural that attention be given to expanded domestic production of feeds for both cow/calf and beef fattening operations. Much attention has been given in Japan to the remarkable successes achieved by the European countries in expanding their animal feedstuffs production (although the Common Agricultural Policy has led to burdensome supplies of them as well as other animal products). Consequently, the purpose of this section is to summarize potential technological changes which Japan can employ to a greater extent, and ones which are in the ofting, to provide insights into how yields might be improved. Then, the next section concentrates on land availability and the potential to increase inventory.

Biotechnology opens the door for rapid increases in plant improvement through recovery of desirable plant genotypes from tissue culture, protoplast fusion, and cloning. Recombinat DNA facilitates the direct manipulation of an organism's genetic material to produce desired characteristics. Although use of rDNA is still in the early stages of development, major advances will be made.

Some plant breeding work being carried out in many countries includes <u>increased photosynthetic efficiency</u> which will probably happen first with corn. It is estimated that an increase of only 1-2 percent in photosynthetic efficiency will double yields. <u>Nitrogen fixing organisms</u> will be used to a greater extent as their biochemistry becomes better known. A major impact in Hokkaido will be <u>corn seed produced by genetic engineering</u> which will result in plants growing to maturity even with early or premature frost. GE will be used to speed up development of new hybrids over conventional plant breeding techniques.

<u>Treatment methods to protect seeds</u> from seed-borne, soil-borne, and mobile organisms are being worked on. In addition, emphasis is being given to treating seeds with plant growth regulators for weed control, and with fertilizers. <u>Plant growth regulators</u> or brassinosteriods, a hormone still in the developmental stage, are expected to have widespread impact on crop yields. They are now being used in Europe to decrease wheat height. <u>Salt tolerance</u> has been bred into more than 50 crops including forage grasses and legumes, cereal crops and oilseeds.

<u>Fertilizer efficiency</u> will be improved greatly. One area involves corn which will provide its own nitrogen and use fertilizer more efficiently. <u>Chemicals</u> for agricultural use are now being synthesized by computer modeling to spot active compounds. This technique will greatly speed up production of new ones. <u>Polyacetylenes</u> are a recently discovered group of plant

chemical insecticides that absorb sunlight and literally burn insects to death.

Some of the more exciting agronomic practices include increased use of double cropping as shorter maturing varieties are developed. Computer controlled planting in which proper depth and spacing is automatically selected to improve uniformity will become common in Japan even on the very small equipment commonly used, due to major cost reduction in small computers. No till (zero-tillage), the practice of reducing the number of tillage passes, will be used to a greater extent where labor and tillage are problems. Forage production using custom prescribed tillage, a dynamic system using information feedback to modify tillage practices throughout the year, will be possible as farmers begin to purchase computers and software becomes available. Forage quality will be improved through plant genetics and management. Included will be more use of legumes in grass swards to decrease nitrogen fertilizer requirements. Hydroponics, a controlled environment agriculture in which plants are grown without soil has considerable potential in densely populated Japan, especially for vegetables. This will release land for grain and forage production. New water conservation practices are also developing rapidly.

Mechanization is an important new area developing in Japan to reduce costs of animal feedstuffs. Innovations include solar bin-buildings to air-dry grain to 20-24 percent moisture. Bale silage bags and numerous other silage storage facilities which provide oxygen-free storage have been developed recently and will be improved in response for more treatment, especially using ammonia to improve protein content. Controlled traffic is a technique in which fixed rows are developed to reduce soil compaction. Engine, draft and tractive efficiency will continue to be improved greatly. This will improve fuel efficiency, extend machinery life, and optimize work rate. Integrated control harvesting is a new system in which sensors on grain combines automatically adjust settings for optimal harvesting, and for information recording and analysis.

Data management and coordination techniques, perhaps the most important aspect in the new scientific revolution, are being developed with bewildering rapidity. A virtual information explosion has taken place in the past five years due to greatly expanded use of computers for data analysis and dissemination of information. Expanded hardware, especially microcomputers, and software will lead to considerable psychological change among farmers and others related to agribusiness vis-a-vis knowledge as a major farm input.

Overall, it can be concluded that major cost reducing and yield enhancing advances will continue to be made in Japanese crop production. Considering that population will only grow at a very small rate in the next quarter century, and Japan is essentially self-sufficient in non-animal feed crop production, the conclusion is an ever growing amount of land will be available for cattle production.

FORAGE PRODUCTION AND LAND UTILIZATION

There are about 1 million ha of land now in forage cultivation and for grazing. Of this, nearly half is in permanent pastures and one-third in annual grasslands, much of which is from diverted land and, to a lesser extent, double cropped paddy land (Table 4.5). About one-fourth is cultivated with crops exclusively or almost exclusively for cattle such as silage or green chop. The average yield of forages on the million acres is about 40 tons per hectare (as fed basis), thus producing about 41 million tons of forage annually.

Rice over-production has become a serious problem in Japan since 1970 and, consequently, there has been official pressure put to divert paddy land to other uses or simply to allow it to remain fallow. In 1970 only about 75,000 hectares of paddy were diverted, of which 20 percent was placed in forage production. By 1975 diverted paddy land had increased to 247,000 ha of which 22 percent was in forage production. By 1983 about 565,000 ha were diverted and 159,000 ha, or 28 percent of the total, were in forage production (Table 4.6). There are about 2.9 million ha of paddy land in Japan so about one-fifth of it is now diverted.

It is anticipated that diverted paddy land will continue to increase in the next quarter century due to modest continued increase in yields, continued decline in per capita consumption, and very low population growth. Scènarios with low and high estimates are now presented to determine additional beef cattle inventory that could be supported by forage production on diverted paddy land. It is estimated that by 1990 between 800,00 and 1,000,000 ha will be diverted and that the diversion will grow to between 1.2 and 1.6 million ha by 2010. These three scenario years--1990, 2000 and 2010 or 5, 15, and 25 years from 1985 will be used throughout the rest of this book as the basis for evaluating the impact of technological change.

The low estimate for proportion in forage production is 30 percent for all three years while the high estimate is 40 percent in 1990 and 50 percent by 2010. It is estimated that under ideal conditions about 5 mature cows and their calves could be supported by forage grown on one ha of paddy land (with purchased

Table 4.5. Forage land and production in Japan, 1970–1983

Item	Year					
	1970	1975	1980	1981	1982	1983
	1,000 Ha					
Total land for grazing and forage cultivation	666	840	1,003	1,026	1,042	1,041
Permanent pastures	279	377	452	461	477	NA
Annual grasslands[a]	194	311	336	337	331	NA
Subtotal	493	688	788	798	808	NA
Land for forage cultivation[b]	173	152	215	228	234	NA
	1,000 Metric Tons					
Average forage yield all land	37	38	38	39	41	40
Total forage production	24,466	32,217	38,490	40,164	43,095	41,274
TDN	2,434	3,208	3,834	4,000	4,292	4,111

Source: MAFF. Statistics on Beef Cattle. Tokyo, 1985.

[a]Includes paddy and other land which is double cropped.

[b]Includes silage and other harvested crops. Three-fourths is corn which is fresh chopped or silage.

Table 4.6. Calculations of additional beef breed cows which could potentially be added to Japan's cattle inventory through land reallocation in 1990, 2000, and 2010

Item	1983	1990	2000	2010
		Paddy Land Reallocation		
Paddy land diverted (Ha)				
Low	565,000	800,000	1,000,000	1,200,000
High	565,000	1,000,000	1,300,000	1,600,000
Proportion in forage production (Pct)				
Low	28	30	30	30
High	28	40	45	50
Paddy land in forage production (Ha)				
Low	159,000	240,000	300,000	360,000
High	159,000	400,000	585,000	800,000
Additional paddy land in forage production over 1983 (Ha)				
Low	—	81,000	141,000	201,000
High	—	241,000	426,000	641,000
Cow/calf equivalents per ha of diverted land (Head)				
Low	—	1.5	1.5	1.5
High	—	3.0	3.0	3.0
Additional cow/calves equivalent on diverted land over 1983 (Head)				
Low	—	121,500	211,500	301,500
High	—	243,000	423,000	603,000
		Marginal Land Reallocation		
Potential marginal land				
Mountains (1,000 Ha)	4,000	4,000	4,000	4,000
Foothills (1,000 Ha)	900	900	900	900
Total (1,000 Ha)	4,900	4,900	4,900	4,900
Reallocation to livestock				
Low (Pct)	—	1	3	5
High (Pct)	—	5	8	15
Reallocated land for livestock (Ha)				
Low (Ha)	—	49,000	147,000	245,000
High (Ha)	—	245,000	392,000	735,000
Cow/calf equivalents per ha of reallocated land (Head)	—	0.25	0.25	0.25
Additional cow/calves on reallocated land over 1983 (Head)				
Low	—	12,250	36,750	61,250
High	—	61,250	98,000	183,750

supplements and additional rice straw as roughage). However, we use estimates of 1.5 and 3.0 cow units supported by one ha as the low and high in the analysis in Table 4.6. The net result is that in 1990 an additional 121,500-243,000 head of cows could be added to national beef cattle inventory over those supported by diverted paddy land in 1983. The total grows to between 301,500-603,000 head by 2010 for the low and high estimates respectively.

Another potential forage source is marginal lands not now used for forage, but which could be brought into production. Two areas exist, mountains and foothills. There are about 4 million ha of mountain land which could potentially be used for spring and fall grazing. Some of this land has been planted to forest for commercial timber production, but is now considered either a very marginal or unprofitable enterprise largely due to lower cost imported wood products.

The second area, foothills, comprises 900,000 ha of moderately sloped areas which are increasingly becoming marginal for timber and crop production (Table 4.6). This land can be used for harvested winter feed such as hay. It appears that by 1990 only about 1-5 percent of the two marginal areas could potentially be brought into production for beef breeds, i.e. between 49,000 and 245,000 ha as a low and high estimate respectively. By 2000 about 147,000-392,000 ha (3-8 percent) might be brought into production, and by 2010, about 245,000-735,000 ha or somewhere between 5 and 15 percent of the total. Given that about 3 ha of the mountain land could carry a cow and a calf, and about 1 ha of the foothills land could be used to feed 1 cow unit, then the low 1990 estimate is for 12,250 cow units while the high estimate is for 61,250 units. The range in 2010 is between 61,250 and 183,750 units.

Overall, it appears that the greatest potential for Japan to expand beef cow inventory is on diverted paddy land. But this diverted land is found nationwide and is quite scattered, even in one locality, which means estimates of its potential use must be quite conservative. The mountain and foothill land is mainly in the Tohoku area, and some institutional difficulty exists since use of forest land for grazing conflicts with other national land use goals.

A third major feedstuffs development possibility is expanded use of rice straw as a source of roughage for cattle. At present rice straw is imported in some areas which have heavy cattle concentrations such as Kyushu, while it is burned in other areas like Kansai where there is a relatively low concentration of cattle. It is expected that as increased amounts of paddy land are diverted in the Kansai area large scale (perhaps 200 head or so) drylot cow/calf operations will develop and will take advantage of the unused rice straw. The potential impact on beef production from adding additional beef cows is discussed in Chapter 9.

SUMMARY AND INFERENCES

Total consumption of animal feeds in Japan has expanded rapidly. Formula feeds now constitute 60 percent of national feed supply on a TDN basis and account for 80 percent of total concentrates used. The greatest increase in formula feeds has taken place in beef cattle, where consumption grew from 15 kg per livestock unit in the 1960-1965 period, to 1,223 kg in 1980.

Imports of coarse grains grew dramatically during the 1970's, reaching about 20.7 million tons in 1984. This is about 98 percent of total coarse grain consumption. The feed use rate in beef cattle (which includes dairy steer fattening) is expected to increase from 8.08 kg of feed per kg of product in 1980 (total national animal basis) to 11.68 kg in 1990 which, along with expanded livestock numbers, and increases in other uses, is expected to result in total consumption of coarse grains increasing to 33 million tons in 1990. Virtually all of that will be from imports.

Technological changes and associated price variations have led to a redundancy in rice paddy land, and a substantial amount of mountain and foothill area. Much of this land could be used for beef production, but reallocation until now has been fairly low. However, it appears that with growing concern about liberalization of beef imports, continued rice yield increases, and few alternative uses for the redundant lands, use of them for beef production, either as grazing or for forage production will grow. Results of an analysis in this chapter indicate that even if only 50 percent of the diverted paddy land were used for beef cattle forage production, beef breed cow numbers and, consequently, beef breed feeder cattle numbers, could be increased by about one-third by 1990 and potentially doubled by 2025. There is a much smaller impact from additional use of marginal lands.

5 Cattle Prices and Grades

Reasonable prices and a good grading system are necessary prerequisites to an expanded cattle industry. In fact, once the mechanisms are in place the profit motive will induce producers to adopt cost reducing technologies. As profits increase there is expanded competition with resultant increased adoption of new technologies and so forth. The purpose of this chapter is to examine cattle prices and grades as background information for production cost estimates and as background for the discussion in later chapters about technological change.

GRADES AND GRADING

There are no grades for live cattle, but a grading system for carcasses has been in effect since 1961. There have been several amendments since that time but they have been minor. There are six categories in the present system. The categories and translations, which vary by source, are given in Table 5.1. A problem which develops is relating the Japanese grades to foreign grades due to different grading locations on the carcass, different emphasis on marbling and perceptions by those in the trade.

Most important in attempting to relate Japanese and foreign grades is location on the carcass. In the Osaka market inspection is at the 7th/8th rib section, while carcasses graded at Tokyo or Kyoto have the cut made between the 5th and 6th ribs. In the United States grading takes place between the 12th and 13th ribs. Marbling decreases and rib eye size increases when moving back along the rib-cage. The U.S. uses its location in the carcass because it is considered to be one of later maturation. Since fat, especially intra-

Table 5.1. Beef carcass grades in Japan

Japanese	Translation	
	Official Statistics	Other
Tokusen	Supreme	Super grade
Gokujyu	Superior	Choice
Jyo	Excellent	1st grade
Chu	Medium	2nd grade
Nami	Common	3rd grade
Togai	Utility	Under-regular grades

muscular (marbling) fat is the last tissue component to be deposited in the animal, the location of the 12th-13th rib is sure to encompass or reflect what has occurred throughout the carcass.

The 5th-6th rib regions (or 7th-8th ribs) where the Japanese carry out their grading is an area of earlier development than that of the 12th-13th rib section, resulting in an apparent higher degree of marbling. For example, if at the 5th-6th rib area there was a slightly abundant degree of marbling (low Prime in the U.S. system) there would only be modest amounts of marbling at the 12th-13th rib area, resulting in a low to average Choice grade.

Meat traders in Japan tend to rank Superior and Excellent as equivalent to U.S. Prime, Medium equal to High and Medium U.S. Choice, and Common as equal to Low Choice or High U.S. Good. Some American meat scientists have argued that these equivalencies tend to downgrade U.S. beef because of differences in carcass location of grading and that, for example, Japanese Excellent is equal to U.S. Medium and High Choice, the Japanese Medium grade is equal to Low Choice, and Common is equivalent to Good.

PRODUCER GOALS AND GRADING RESULTS

One common link between producers and all participants in the marketing chain is a belief that Wagyu beef should be as heavily marbled as possible. Consequently, the whole feeding strategy of Wagyu is aimed at this one, fundamental over-riding goal. It is the reason why pedigree is so important in feeder calf purchase decisions, and it accounts for what appears to many foreigners as irrational economic behavior on the part of cattle feeders. But, on reflection, the behavior is reminiscent of the approach to cattle feeding in the United States Midwest in the early 1950's when the signal to producers was "pour the corn to them and shoot for High Prime." Another important consideration is inertia in production and marketing customs that is difficult to overcome.

Roughly speaking, there are two classes of Waygu feeders; a rather small group trying for Supreme and Superior and a second group targeting for the middle grades but hoping some cattle will reach the high grades. Both groups drag out the feeding period in the mistaken belief that a longer feeding period, and thus a slower gain will result in a higher degree of marbling than a short feeding period. The result is Wagyu cattle are on feed for about 18 months and slaughtered at around 27-28 months of age weighing about 630-635 kg.

Despite much discussion about the ability of Wagyu to marble, it turns out that less than 2.0 percent of all steers or heifers actually make the Supreme grade (Table 5.2 and Appendix 5.1). Also, only about another 4-5 percent make the second grade, Superior, and just about 10-20 percent grade Excellent, the third grade. This leaves the vast majority in the three lowest grades; Medium, Common and Utility.

Producers do recognize that only a small proportion of Wagyu will make the highest grades. But, when all factors are taken into account, even though virtually all producers target their entire calf purchase, feeding strategy, and finished animal marketing on reaching as high a grade as possible, in reality few of them succeed in the gamble, and most just turn a large quantity of American grain into fat which is subsequently trimmed off and disposed of before the meat is presented to the consumer.

The proportion of Wagyu graded has increased steadily, from just 23.6 percent in 1972 to 45.7 percent in 1980 and 56.9 percent in 1984 (Table 5.3). The expansion in numbers graded reflects longer distances transported due to an improved highway system in Japan, rapid development of regional packing plants, closing of many small municipal abattoirs and, of course, growth of commercial cattle feeders intent on receiving the highest possible price for their cattle.

Dairy cattle producers, as described in earlier chapters, are much more pragmatic about their feeding program than Wagyu producers. Their cattle are on feed for a much shorter period (about 12-13 months versus about 18 for Wagyu) although the sale weight is heavy

Table 5.2. Number and grade of Wagyu steers slaughtered and graded in Japan, 1972-1984

Year	Total graded	Total steers slaughtered	Proportion graded	Grade: Supreme Tokusen	Grade: Superior Gokujyu	Grade: Excellent Jyo	Grade: Medium Chu	Grade: Common Nami	Grade: Utility Togai
	---- Head ----		---- Percent ----						
1972	54,687	219,391	24.9	0.8	5.5	38.4	50.9	3.8	0.6
1973	43,874	142,788	30.7	1.1	6.1	37.7	50.7	3.7	0.8
1974	60,288	169,170	35.6	1.2	4.7	27.9	56.8	8.6	0.8
1975	78,812	212,807	37.0	1.8	6.3	27.5	54.6	8.9	0.9
1976	78,433	191,967	40.9	1.8	5.8	25.0	54.6	11.8	1.0
1977	89,788	203,129	44.2	1.3	4.5	21.8	54.6	16.8	1.0
1978	111,359	236,604	47.1	1.2	3.8	19.0	56.7	18.4	0.9
1979	103,959	217,326	47.8	1.7	5.1	24.9	53.2	14.3	0.8
1980	102,843	202,735	50.7	2.0	5.7	25.9	50.8	14.7	0.9
1981	119,500	213,544	56.0	1.7	4.9	21.0	51.8	19.7	0.9
1982	132,611	220,085	60.3	1.6	4.6	19.0	52.6	21.1	0.9
1983	151,536	240,280	63.1	1.6	4.4	17.9	50.2	24.7	1.1
1984	180,575	269,900	66.9	1.4	3.7	14.3	46.6	32.8	1.2

Source: Derived from MAFF. <u>The Meat Statistics in Japan</u>. Tokyo, 1985.

Table 5.3. Comparison of total cattle slaughtered with those graded in Japan, 1972-1984

	Total Wagyu			Total dairy		
Year	Slaughtered	Graded	Proportion	Slaughtered	Graded	Proportion
	- - - - Head - - -		- Percent -	- - - - Head - - -		- Percent -
1972	524,951	123,647	23.6	561,889	132,588	23.6
1973	318,167	87,828	27.6	487,734	114,042	23.4
1974	320,395	103,238	32.2	662,905	177,626	26.8
1975	406,121	132,288	32.6	736,968	205,679	27.9
1976	400,598	144,252	36.0	546,605	176,910	32.4
1977	439,863	171,184	38.9	666,715	243,549	36.5
1978	491,930	205,302	41.7	706,204	269,591	38.2
1979	432,519	186,405	43.1	753,415	302,638	40.2
1980	374,147	171,039	45.7	812,892	341,601	42.0
1981	368,764	182,467	49.5	952,925	419,322	44.0
1982	387,636	199,678	51.5	957,399	439,564	45.9
1983	445,121	238,854	53.7	933,214	434,570	46.6
1984	554,057	315,457	56.9	938,523	453,567	48.3

Source: Adapted from MAFF. The Meat Statistics in Japan. Tokyo, 1985.

(665 kg) compared with United States feedlots (sale at about 475 kg). Dairy cattle feeders do attempt to obtain the highest possible grade but. of course, the final average grade of dairy cattle is considerably lower than Wagyu.

Total dairy cattle slaughter has increased continuously, from 562 thousand head in 1972 up to about 950 thousand head annually since 1980. The proportion which are graded has grown at a rapid rate, nearly doubling from 23.6 percent in 1972, to 48.3 percent in 1984 (Table 5.3). This growth in grading has paralleled the development of Japan's commercial feedlot industry.

As would be expected, dairy animals generally do not reach the Supreme or Superior grade. In fact, only about a half to 1 percent of dairy steers are stamped Excellent while about 50 percent grade Medium (Table 5.4). About half grade Common. Some female dairy cattle (about 0.2-0.4 percent) reach Superior, 1-3 percent grade Excellent. and about 20-25 percent Medium (Appendix 5.2).

There has been considerable criticism of the grading system from many diverse groups. As a result, an ad hoc advisory committee has prepared a report for review by MAFF. Their main points are (1) the need for yield grading, (2) desirability of less emphasis on marbling even to the extent that the highest grade, Supreme, would have lower marbling requirements, (3) the same number of grades should be maintained, (4) dairy and beef breeds should continue to fall under the same grading standards, and (5) the location of grading should be standardized between the 6th and 7th rib.

CATTLE PRICES

Cattle prices are subject to a whole series of policies designed to protect consumers, producers and to stimulate production. The *modus operandi* is for government to set floor and ceiling prices to ensure they only fluctuate within a given range (Table 5.5). The policies act as a guarantee to producers and, along with subsidies that vary widely depending on the region of the country, time of year, and type of cattle, help promote domestic production. The core of the policy is that, when market prices drop near the floor price, beef imports are restricted which then causes prices to rise. When beef prices approach or exceed the ceiling price, additional quantities of imported beef are released on the market. The policies are not rigid, but rather are used as guides.

There is considerable price difference between grades of beef, and they help explain why cattle feed-

Table 5.4. Number and grade of dairy steers slaughtered and graded in Japan, 1972-1984

Year	Total dairy cattle graded			Grade of steers					
	Steers	Graded	Proportion	Supreme Tokusen	Superior Gokujyu	Excellent Jyo	Medium Chu	Common Nami	Utility Togai
	- - - - Head	- - - -	- Percent -	- - - - - - - -	- - - - - -	- - - Percent	- - - - -	- - - - -	- - - -
1972	132,585	47,411	35.8	—	0	1.0	49.8	46.4	2.8
1973	114,042	47,650	41.8	—	0	1.6	55.2	40.2	3.0
1974	177,625	94,569	53.2	0	0	0.6	41.0	52.1	6.3
1975	205,679	101,961	49.6	0	0	0.7	40.5	53.5	5.3
1976	176,910	81,081	45.8	0	—	0.9	44.2	51.1	3.8
1977	243,549	121,560	49.9	0	0	0.5	46.6	48.7	4.2
1978	269,591	142,595	52.9	—	0	0.7	49.7	46.2	3.4
1979	302,638	153,409	50.7	0	0	1.1	54.2	41.2	3.5
1980	341,601	178,591	52.3	—	0	1.0	52.9	42.5	3.6
1981	419,322	216,086	51.5	0	0	0.6	51.9	43.7	3.8
1982	439,564	228,137	51.9	0	0	0.4	53.1	43.4	3.1
1983	434,570	226,637	51.3	0	0	0.4	49.6	45.4	4.6
1984	455,567	237,206	52.1	0	0	0.4	43.9	51.3	4.4

Source: MAFF. The Meat Statistics in Japan. Tokyo, 1985.

Table 5.5. Floor and ceiling carcass price, average of central wholesale meat markets, Japan, 1975–1984

Fiscal Year[a]	Wagyu steers		Other steers	
	Floor price	Ceiling price	Floor price	Ceiling price
	- - - - - -	- - - Yen per kg	- - -	- - - - - -
1975	1,143	1,518	930	1,236
1976	1,240	1,647	1,009	1,341
1977	1,303	1,730	1,061	1,408
1978	1,303	1,730	1,061	1,408
1979	1,303	1,730	1,061	1,408
1980	1,357	1,763	1,105	1,435
1981	1,399	1,817	1,118	1,452
1982	1,400	1,820	1,120	1,455
1983	1,400	1,820	1,120	1,455
1984	1,400	1,820	1,120	1,455

Source: MAFF. The Meat Statistics in Japan. Tokyo, 1985.

[a]April through March.

ers are willing to gamble on reaching the high grades. Over the period 1982-1984, those Wagyu feeders fortunate (or skillful) enough to reach the Supreme grade (less than 2 percent of cattle graded) received a price premium 63 percent above the Medium grade. Superior fetched 40 percent more than Medium (but only 4-5 percent of cattle reached this grade), while Excellent resulted in a 20 percent premium (Table 5.6). The cost of only making Common was a 21 percent discount below Medium. The weighted average price for all grades was 8 percent above Medium.

Beef carcass prices (Excellent grade) have remained quite stable since 1976. But, in U.S. dollar terms, because of fluctuations in exchange rates, the value of Medium grade increased from $5.71 per kg in 1976, to $8.31 in 1979 (Table 5.7). It has fluctuated between $7.32 and $8.15 since then, again mainly due to exchange rate fluctuations. In the meantime, beef prices in the United States increased from $1.34 per kg in 1976 to $2.30 in 1980 but declined to $2.20 in 1984. Thus, roughly comparable prices show domestically produced beef to have been about 3.5 times that in the U.S. in the past seven years.

Prices (in yen) of Wagyu calves (feeder steers) increased about 35 percent from 1978-1983 while finished steers went up 12 percent (Table 5.8). The rapid increase in calf prices while finished cattle prices have remained quite stable is largely due to efficiencies that have resulted from commercialization of cattle feeding through development of large scale lots. Production costs of these lots have declined and, consequently, they have been able to bid up the price of feeder calves. Government has floor and ceiling prices on slaughter cattle, but calf prices are set by the market through supply and demand forces. Feeder steer prices averaged about 19 percent more than fed steers prices from 1981-1983.

The average price of Medium grade dairy steer carcasses was ¥1,295 per kg from 1982-1984 (Table 5.9). Prices in 1984 of Common grade steer carcasses were 88 percent of Medium while Utility was 62 percent of Medium. A 22 percent premium was achieved for reaching Excellent. The average price for calves entering feedlots was ¥732 per kg from 1980-1983 compared with ¥727 for finished steers (Table 5.10). Price of calves is thus nearly equal to finished steers.

Price on Excellent and Medium grade dairy carcasses was 27 percent less than for the same grade Wagyu carcasses during that period (Table 5.11) indicating the great amount of product differentiation afforded to the Wagyu name since the same grading standards are used for both types of cattle.

Structural changes in Japan's meat industry have led to considerable pricing problems. The 10 public

Table 5.6. Wagyu slaughter steer carcass price and index of prices by grade, Tokyo wholesale market, 1972-1984

Year	Grade						Weighted average
	Supreme	Superior	Excellent	Medium	Common	Utility	
	Yen/kg						
1972	1,198	1,066	935	848	716	612	868
1973	1,841	1,578	1,419	1,283	1,042	817	1,383
1974	1,939	1,697	1,445	1,153	886	710	1,345
1975	2,092	1,828	1,640	1,433	1,235	996	1,573
1976	2,601	2,254	1,986	1,696	1,367	986	1,878
1977	2,764	2,344	2,007	1,618	1,304	980	1,819
1978	2,749	2,323	1,938	1,584	1,356	1,038	1,766
1979	2,718	2,363	2,056	1,819	1,575	1,175	1,967
1980	2,796	2,460	2,161	1,851	1,474	1,092	2,046
1981	2,792	2,458	2,142	1,791	1,434	1,084	1,946
1982	2,841	2,487	2,155	1,823	1,467	1,084	1,957
1983	2,898	2,515	2,168	1,786	1,405	1,004	1,953
1984	3,019	2,522	2,143	1,749	1,384	960	1,894
Average 1982-1984	2,919	2,508	2,155	1,786	1,419	1,016	1,929
	Index Medium = 100						
1972	141	126	110	100	84	72	106
1973	143	123	111	100	81	63	108
1974	168	147	125	100	77	62	117
1975	146	128	114	100	86	70	110
1976	153	133	117	100	81	58	111
1977	171	145	124	100	81	61	112
1978	174	147	122	100	86	66	111
1979	149	130	113	100	87	65	108
1980	151	133	117	100	80	59	111
1981	156	137	137	100	80	61	109
1982	156	136	118	100	80	59	107
1983	162	141	121	100	79	56	109
1984	173	144	123	100	79	55	107
Average 1982-1984	163	140	120	100	79	57	108

Source: Derived from LIPC. Japan's Beef Figures. Tokyo, selected monthly issues.

Table 5.7. Comparison of beef carcass prices in Japan and the United States, 1973-1984

Year	Japan Medium grade Wagyu steers			United States Choice steer beef[a]		Japan as a proportion of the U.S.
	Per kg	Exchange rate	Per kg	Per lb	Per kg	
	- Yen -		- $ US -	- - - - $ US - - - -		- - Ratio - -
1973	1,283	271	4.73	0.67	1.47	3.2
1974	1,153	292	3.95	0.67	1.47	2.7
1975	1,433	297	4.82	0.73	1.61	3.0
1976	1,696	297	5.71	0.61	1.34	4.3
1977	1,618	271	5.97	0.63	1.38	4.3
1978	1,584	210	7.54	0.81	1.78	4.2
1979	1,819	219	8.31	1.02	2.24	3.7
1980	1,851	227	8.15	1.04	2.30	3.5
1981	1,791	221	8.10	1.00	2.20	3.7
1982	1,823	249	7.32	1.02	2.24	3.3
1983	1,786	238	7.50	0.98	2.16	3.5
1984	1,749	240	7.29	1.00	2.20	3.3

Source: Japanese price from Table 5.6. United States prices from USDA, *Livestock and Meat Statistics*, various issues and USDA/AMS.

[a]Choice 700-800 lb steer carcasses, carlot basis, Central U.S.

Table 5.8. Sale weights and calculated prices of Wagyu calves and steers in Japan, 1972-1983

Year	Sale weight	Price Per head	Price Per kg	Price[a] Per head	Price[a] Per kg
		---- Yen ----		---- $ US ----	
		Feeder steers			
1972	300.5	137,825	459	454.90	1.51
1973	295.1	150,691	511	554.00	1.88
1974	292.4	202,977	694	695.10	2.38
1975	282.7	288,860	1,022	972.60	3.44
1976	275.5	228,204	828	768.36	2.79
1977	268.4	224,864	838	829.76	3.09
1978	277.4	265,739	958	1,265.42	4.56
1979	291.9	289,345	991	1,321.21	4.53
1980	286.6	296,013	1,033	1,304.02	4.55
1981	282.0	350,720	1,244	1,586.97	5.63
1982	286.0	387,959	1,357	1,558.07	5.45
1983	286.5	363,573	1,269	1,527.62	5.33
Average 1981-1983	284.8	367,417	1,290	1,557.55	5.47
		Slaughter steers			
1972	532.6	262,209	492	865.40	1.62
1973	542.9	350,480	646	1,288.50	2.38
1974	565.7	453,962	802	1,554.70	2.75
1975	587.9	458,622	780	1,544.20	2.63
1976	564.4	567,567	1,006	1,911.00	3.39
1977	581.8	602,751	1,036	2,224.17	3.82
1978	597.4	577,639	967	2,750.66	4.60
1979	605.8	620,302	1,024	2,832.43	4.68
1980	613.1	690,402	1,126	3,041.42	4.96
1981	630.3	696,270	1,105	3,150.54	5.00
1982	639.1	690,714	1,081	2,773.95	4.34
1983	632.9	668,109	1,056	2,807.18	4.44
Average 1981-1983	634.1	685,031	1,080	2,910.56	4.59
	Feeder steers as a proportion of slaughter steers				
	------ Percent -----				
1972	56.4	52.6	93.3		
1973	54.4	43.0	79.1		
1974	51.7	44.7	86.5		
1975	48.1	63.0	131.0		
1976	48.8	40.2	80.3		
1977	46.1	37.3	80.9		
1978	46.4	46.0	99.1		
1979	48.2	46.6	96.8		
1980	46.7	42.9	97.7		
1981	44.7	50.3	112.6		
1982	44.8	56.2	125.5		
1983	45.3	54.4	120.2		
Average 1981-1983	44.9	53.6	119.4		

Source: Derived from MAFF. The Meat Statistics in Japan. Tokyo, various issues.

[a]Exchange rates in Appendix 1.9.

Table 5.9. Dairy steer carcass price and index of prices by grade, Tokyo wholesale market, 1975–1984

Year	Grade						Weighted average
	Supreme Tokusen	Superior Gokujyu	Excellent Jyo	Medium Chu	Common Nami	Utility Togai	
	Yen/kg						
1975	—	1,691	1,381	1,234	1,106	849	1,157
1976	—	2,261	1,661	1,358	1,200	833	1,268
1977	—	1,920	1,514	1,265	1,158	834	1,199
1978	—	1,501	1,515	1,276	1,183	817	1,219
1979	—	2,103	1,636	1,451	1,353	1,004	1,400
1980	—	1,998	1,614	1,329	1,212	888	1,278
1981	—	2,174	1,502	1,245	1,133	833	1,196
1982	—	1,853	1,565	1,304	1,183	870	1,258
1983	—	—	1,537	1,298	1,146	829	1,234
1984	—	—	1,567	1,284	1,134	794	1,182
Average 1982–1984	—	—	1,556	1,295	1,154	831	1,225
	Index Medium = 100						
1975	—	137	112	100	90	69	94
1976	—	166	119	100	88	61	93
1977	—	152	120	100	92	66	95
1978	—	118	119	100	93	64	96
1979	—	145	113	100	93	69	96
1980	—	150	121	100	91	67	96
1981	—	175	121	100	91	67	96
1982	—	142	120	100	91	67	96
1983	—	—	118	100	89	64	95
1984	—	—	122	100	88	62	92
Average 1982–1984	—	—	120	100	89	64	94

Source: LIPC. Japan's Beef Figures. Tokyo, selected monthly issues.

Table 5.10. Sale weights and calculated prices of dairy calves and steers, in Japan, 1976–1983

	Dairy calves			Finished dairy steers			Price as a proportion of finished steers	
	Sale weight	Price		Sale weight	Price			
		Per head	Per kilo		Per head	Per kilo	Per head	Per kilo
	- Kg -	- - - - Yen	- - - -	- Kg -	- - - - Yen	- - - -	- - Percent	- -
1976	268.1	120,188	448.3	573.8	445,683	776.7	27.0	57.7
1977	254.6	146,801	576.6	594.2	431,382	726.0	34.0	79.4
1978	277.7	155,508	560.0	623.9	435,479	698.0	35.7	80.2
1979	276.7	159,641	576.9	631.5	482,224	763.6	33.1	75.6
1980	260.2	184,082	707.5	643.7	524,218	814.4	35.1	86.9
1981	265.7	220,606	830.3	662.4	469,522	708.8	47.0	117.1
1982	268.9	189,686	705.4	664.9	484,624	728.9	39.1	96.8
1983	267.8	176,449	658.9	667.3	495,379	742.4	35.6	88.8
Average 1980–1983	267.5	195,580	731.5	664.9	483,175	726.7	40.6	100.9

Source: LIPC. Japan's Beef Figures. Tokyo, selected monthly issues.

Table 5.11. Price differences between Wagyu and dairy slaughter steers by grade, Japan, 1982-1984

	Grade					
	Supreme Tokusen	Superior Gokujyu	Excellent Jyo	Medium Chu	Common Nami	Utility Togai
	- - - - -	- - - - -	- - Yen/kg	- - - - -	- - - - -	- - - - -
Average, Wagyu	2,919	2,508	2,155	1,786	1,419	1,016
Average, Dairy	--	--	1,556	1,295	1,154	831
Difference	--	--	599	491	265	185
Dairy as a percent of Wagyu	--	--	72	73	81	82

Source: Wagyu prices from Table 5.6. Dairy prices from Table 5.9.

wholesale markets--the source from which price reports are derived--only accounted for about 20-25 percent of all cattle slaughtered in 1984. There is continued expansion of direct sales, with most of this as boxed beef. The Kawasaki Meat Trading Center near Tokyo was expected to fulfill a price reporting role, but has not proved to be an effective instrument. As a result there is no good pricing system for boxed beef.

SUMMARY AND CONSIDERATIONS

Beef cattle feeders in Japan, and Wagyu feeders in particular, are dominated by an all-encompassing goal of having their cattle reach as high a grade as possible. Several tactics are used depending on the particular type of cattle being fed, but in all cases the cattle are carried to very heavy weights. The trade is used to these type carcasses (probably a holdover from the days when beef was a joint product of cattle as a draft animal) and does discount light carcasses. In effect there is a premium paid for maturity. But, as consumer preferences have shifted to much more differentiated products there is question about the long feeding tactic.

Long feeding periods result in reduced average daily gains and slow turnover in feedlots. The net result is relatively high cost of gain and high break-even points. Custom among both producers and wholesalers leads to the long anti-economic feeding periods; as a result a major policy question is how to overcome the inertia. One way is to change the grading system to favor younger, lighter carcasses and to stress yield as well as marbling. An ad hoc advisory committee has been discussing grade changes and it appears that a new system may be approved in 1986 or 1987. Adoption of a new grading system, which would lead to reduced time on feed, could easily lead to a 3-5 percent cost reduction in Wagyu cattle by 1990 and 5-10 percent by 2000. Comparable figures for dairy are about 1-3 percent by 1990 and 4-6 percent by 2000.

Carcass prices in Japan for Excellent (Jyo) grade are about 3.5 times Choice carcass grade prices in the United States. The difficulties in determining grade equivalencies between foreign and Japanese carcasses were pointed out in this chapter. If the Japanese Medium (Chu) is equal to U.S. High and Medium Choice, then the equivalent of Choice beef produced in Japan is about 3 times Choice beef in the U.S. Given that about 30 percent of Japan's beef is imported, and that most of it is imported from Australia and is significantly lower priced than U.S. beef, it becomes clear why the _average_ price of beef in Japan is only about 2-2 ½ times that in the United States. It is also the reason why hamburgers at fast food chains are about the same price in Japan as the United States and why a majority of beef cuts are equivalent or only slightly more than in the United States. In effect, it is a mistake to simply compare Choice U.S. sirloin steak with highly marbled "shimofuri" thin sliced product.

The need to understand product differentiation is highlighted by evaluation of grading results presented in this chapter. Statistics show that about 50 percent of beef graded is Medium (Chu) and about 33 percent (weighted between dairy and Wagyu) is Common (Nami). About 12 percent of the beef is from Utility (Nami). This means that nearly 90 percent of the beef produced in Japan falls into the "popular" grades. Furthermore, when beef going into low priced cuts from high priced carcasses is added to the total, it is clear that the so-called "very high quality" beef occupies only a very restricted part of the market.

Overall, it is apparent that Japan has been quite successful in stabilizing beef prices and, if they were deflated, would have a slight downward trend. In the meantime EEC prices have trended upward and U.S. prices have fluctuated although they were fairly stable in the early 1980's. The Japanese government is concerned about the means to maintain stable prices yet increase domestic production. The only solution appears to be reducing production cost. The rationale is lower cost would lead to increased short-term profits which in turn would provide incentive to expand production.

The policy question which develops from examination of the supply and demand side regarding grades and prices is the extent to which government should continue to shelter producers through the awkward system of subsidies and the price stabilization scheme. An alternative is to introduce a new grading system combined with rapid provision of supply side technologies. This aspect is considered next as we compare costs and returns between producers in Japan and the United States. Then Chapter 7 follows with a review of the various technologies now available, and which will become available, to Japanese beef producers over the next quarter of a century.

6 Production Costs and Returns

Cost and returns estimates for both Japanese and U.S. beef breed cow/calf operators, beef breed cattle feedlots, and for dairy breed feeding to slaughter weights are presented in this chapter. There are, of course, considerable differences between operations within a country, but the budgets do serve the major function which is to provide comparative estimates of Japanese and American beef production cost and the reason behind them. The cost estimates for Japan serve as a base from which economic analyses of the potential impact from technological changes are carried out in Chapters 8 and 9. In effect, a second major purpose of this chapter, as well as the background information developed in previous chapters and a discussion of technological changes in the next chapter, is to answer the question: How much can beef production cost be reduced in Japan?

BEEF BREED PRODUCTION CHARACTERISTICS

About 30 percent of Japan's beef production is from beef type cattle, with the nearly exclusive breed being Wagyu. Overall, there were only 315 thousand farms with beef cattle in 1984 compared with 1.4 million farms in 1965. Beef cattle inventory (government statistics include steers from dairy breeds) increased from 1.8 million head in 1965 to 2.6 million head in 1984. The number of beef cattle per farm increased from an average of 1.3 head in 1965 to 8.2 in 1984.

The trend toward fewer and larger farms with cattle has continued during recent years and all indications are that the decline will continue. In 1975 there were 3,545 farms with 50 head of beef cattle or more; by 1984 they had more than doubled to 7,880 farms. As of 1975 about 0.7 percent of all beef cattle were found on these farms; by 1984 about 2.7 percent of the total were located on them. In the meantime, cattle inventory of farms with 50 head of more increased from 14.2 percent of total beef cattle inventory to 31.9 percent. Virtually all of this increase is in feedlot development.

Feeder cattle in Japan are primarily produced on small operations, ones with 1-4 head. Farms of this size decreased from 84.8 percent of all farms with beef cattle in 1975 to 70.0 percent in 1984. Inventory of Japanese farms with 4 head or less of beef cattle declined from 41.8 percent of total beef cattle in 1975 to just 21.8 percent nine years later in 1984.

Crossbreeding is not used on beef operations and an inordinate amount of attention is given to pedigree, at least from Western eyes. Also, commercial grade feeder cattle production displays characteristics similar to a U.S. purebred operation except on a much smaller scale. The cows on small operations are maintained in close confinement and feed is brought to them. Over 90 percent of beef cows are bred using artificial insemination.

There is no stocker or backgrounding (i.e. a growing period just prior to entering the feedlot) phase for beef breed cattle per se except in rare instances, since pasture land is quite limited. A restricted amount of wheat land is available, but due to lack of fences, distances involved, and other operational difficulties, there is virtually no use made of it for grazing. Consequently, cattle, mainly steers, go directly from the auction market to a confinement feeding facility.

The traditional way to feed Wagyu cattle (mainly virgin heifers) has been to keep them in small barns holding just a few head. There are still many of these very small operations but they are rapidly disappearing. Their feeding system has been to keep the animal locked in a stall, have the manure cleaned at regular intervals, and bedding changed frequently. The rations have historically been prepared in individual buckets, often by making a gruel with warm water to improve palatability and digestibility. Cattle have been provided careful individual attention giving rise to a multitude of largely fictitious stories about production of "Kobe beef" which had additional marbling added by being heavily brushed, massaged, fed beer, and through religious rites.

The modern lots feeding Wagyu found today are relatively large scale operations, from 500 to 5,000 head or more. Investment in facilities is usually extensive, even in the rather mild climate found in Southern Japan. One reason for this is the very long period cattle are on feed and great concern about the animal's comfort. There are some slatted floor facilities with manure falling into a sump, but generally the facility is a large barn divided into long rows of concrete floor pens each holding 10-20 head, and with bedding. Cattle are generally only taken outside when bedding is changed--in fact, the objective is to restrict movement as much as possible as a strategy to improve marbling. There are essentially no open dirt lots. In some cases feeding in the commercial operations is carried out using self-mixing feedwagons or trucks, while in other cases it is done by hand. Great care is taken in manure disposal as pollution is a major concern in most areas where cattle are fed.

Wagyu cattle feeders are very quality conscious. That overriding goal, as discussed in the last chapter, is really taken to an extreme in both traditional and commercial lots as cattle gains are deliberately slowed down in an effort to achieve the highest possible degree of marbling by (1) reducing the amount of feed provided at the typical twice daily feeding so the cattle clean up all feed in just a few hours and/or (2) by maintaining cattle on a high roughage diet until the last phase of the feeding period at which time an 80 percent concentrate ration is fed. Even in the last phase many feeders regulate quantity so the cattle are without feed for limited periods.

BEEF BREED COW/CALF COSTS AND RETURNS

Regional and national cost and returns production budgets are now presented. These type budgets rather than ones for individual operations mask some differences between operations. Consequently, they provide only limited opportunity to understand individual differences in component costs. Nevertheless, national budgets are useful to obtain a basic understanding of differences between countries.

A general quantitative description of Wagyu cow/calf operations, typified in MAFF cost and returns budgets for Kyushu, Japan in 1980 and 1983 is given in Appendix 6.1. Explanation of labor use for the two periods is provided in Appendix 6.2. Both tables reveal the significant changes which are taking place in Japan due to economies of size. For example, total hours per Wagyu calf produced declined 20 percent during the 3 year period. Budget data, translated directly from the MAFF publication on production costs, is given in Appendix 6.3 with a note by the side of each item to show the grouping of items in the summary provided in Table 6.1.

Budgets for cow/calf operations in the United States are provided in Appendix 6.4. There are, of course, many regions in the U.S. so selection of the

Table 6.1. Cost and returns comparisons for cow/calf enterprises, Japan and the United States, 1980 and 1983

Item	Kagoshima, Japan 1980	Kagoshima, Japan 1983	Great Plains United States 1980	Great Plains United States 1983
	Data as Published			
	- - Yen per calf - -		- Dollars per cow -	
Cash receipts	391,484	229,577	315.87	252.56
Cash expenses (variable)				
Feed	190,942	207,670	96.06	108.82
Nonfeed	46,793	46,552	56.14	64.90
Subtotal	237,735	254,222	152.20	173.72
Manure credit	41,069	55,172	—	—
Total variable	196,666	199,050	152.20	173.72
Cash expenses (fixed)				
General farm overhead	—	—		—
Taxes and insurance	—	—		—
Interest	30,844	34,826		44.76
Land rent	15,062	22,795		—
Total fixed	45,906	57,621	NA	44.76
Total cash	242,572	256,671	NA	218.48
Economic (total) costs				
Variable	196,666	199,050		173.72
Fixed	45,906	57,621		44.76
Capital replacement (depreciation)	60,760	67,768		59.61
Allocated returns				
Operating capital	—	—		—
Other nonland capital	—	—		—
Land	—	—		—
Unpaid labor	141,703	132,719		67.10
Total economic[a]	445,035	457,158	NA	345.19
	Expenses and Net Income Per Cow			
	- - - - - - - - - - - - Dollars - - - - - - - - - -			
Expenses[b]				
Variable	727.75	719.26	152.20	173.72
Total cash	897.62	927.47	NA	218.48
Total economic	1,646.83	1,651.92	NA	345.19
Cash receipts	1,448.66	829.56	315.87	252.56
Net income per cow				
Above variable	720.91	110.30	163.67	78.84
Above cash	551.04	-97.91	NA	38.08
Above total	-198.17	-822.36	NA	92.63

Source: Appendixes 6.3 and 6.4. United States data for 1980 from USDA Livestock and Poultry: Outlook and Situation, LPS-15, March, 1985.

[a]Not all items included from U.S. budget (Appendix 6.4) to provide comparability.

[b]Japanese data adjusted by 84 percent weaned calf crop and exchange rate of ¥227 in 1980, and 86 percent weaned calf crop and exchange rate ¥238 in 1983.

Great Plains is somewhat arbitrary although examination of costs and returns for the various regions reveals considerable similarity between them, except for the Southeast which has much higher costs. Not all items from the section called "fixed economic costs" in Appendix 6.4 are included in Table 6.1 as the objective is to provide comparability. The Japanese budgets do not include several items which are found in the U.S. budgets.

The summary budgets in Table 6.1 show that variable costs on Wagyu operations are about 4 times as high as in the U.S., $719 per cow versus $174 in 1983. Total cash costs, which do not include labor, are also about 4 times those of the United States, $927 versus $218. Total cash and noncash (economic)costs are $1,652 versus $345, i.e. Japan's costs on a total basis are about 4.8 times those in the United States. The total for Japan is very high because labor, which is quite high on a per cow basis due to small scale operations, is included as a noncash cost. Comparison of expenses for 1980 and 1983 reveal that expenses per cow increased only slightly in Japan (due to low inflation and actual cost declines in some items) while expenses increased considerably in the United States.

SLAUGHTER CATTLE PRODUCTION COSTS AND RETURNS

A comparison of production costs and returns between Japan and the United States for feeding cattle from calves to slaughter weight is now provided. A rather small Kyushu feedlot (16 head) is quantitatively described in Appendix 6.5. There were 2.6 workers per farm in 1980, but the average had declined to 2.4 in 1983. The average time on feed was about 17 months in each time period. Hours spent declined 22 percent from 1980 to 1983 (Appendix 6.6).

Details on average production costs for feeding Wagyu in Kyushu are provided in Appendix 6.7 and summarized in Table 6.2. Variable cash costs increased slightly during 1980 to 1983, from $3.78 per kg to $3.89. Total costs (cash and noncash) increased slightly, from $4.49 to $4.54 per kg. Income declined so the net result is returns above variable costs declined from $1.08 to $0.49 per kg, while returns above total costs were $0.37 and $-0.16 for the two periods respectively. Part of the reason for the apparent lower returns in 1983 is a stronger dollar, valued at 238 yen compared with 227 in 1980. Exchange rate fluctuations both between years and within years result in considerable differences in comparisons.

Table 6.2. Cost and returns comparisons for Japanese Wagyu feedlots and three types of United States feedlots, 1980 and 1983

Item	Kyushu, Japan 1980	Kyushu, Japan 1983	United States, 1983 Farmer	United States, 1983 Commercial	United States, 1983 All sizes
			Data as Published		
	- - Yen per steer - -		- - - - - Dollars per cwt - - - - -		
Cash receipts	670,061	655,176	62.03	62.15	62.12
Cash expenses (variable)					
Feed	221,631	241,516	18.18	20.92	20.31
Nonfeed	17,220	21,835	5.10	4.85	4.91
Feeder cattle	302,227	346,390	35.21	38.30	37.62
Subtotal	541,078	609,741	58.49	64.07	63.29
Manure credit	20,525	27,908	0.11	0.10	0.10
Total variable	520,553	581,833	58.38	63.97	62.74
Cash expenses (fixed)					
General farm overhead	—	—	—	—	—
Taxes and insurance	—	—	—	—	—
Interest	18,710	18,735	4.64	4.08	4.22
Land rent	4,345	5,627	—	—	—
Hired management	0.00	0.00	0.00	0.18	0.14
Total fixed	23,055	24,362	4.64	4.26	4.36
Total cash	543,608	606,195	63.02	68.23	67.10
Economic (total) costs					
Variable	520,553	581,833	58.38	63.97	62.74
Fixed	23,055	24,362	4.64	4.26	4.36
Capital replacement (depreciation)	12,615	14,036	3.62	0.52	1.21
Allocated returns					
Operating capital	—	—	—	—	
Other nonland capital	—	—	—	—	
Land		—	—	—	—
Unpaid labor	62,293	60,173	2.37	0.00	0.52
Total economic[a]	618,516	680,404	69.01	68.75	68.83
			Per Kilo of Live Animal Sold		
	- - - - - - - - - - - - - Dollars - - - - - - - - - - - - -				
Expenses[b]					
Variable	3.78	3.89	1.28	1.41	1.38
Total cash	3.95	4.05	1.39	1.50	1.48
Total economic	4.49	4.54	1.52	1.51	1.51
Cash receipts	4.86	4.38	1.36	1.37	1.37
Net income per kg					
Above variable	1.08	0.49	0.08	-0.04	-0.01
Above cash	0.91	0.33	-0.03	-0.13	-0.11
Above total	0.37	-0.16	-0.16	-0.14	-0.14

Source: Appendixes 6.7 and 6.8.

[a]Not all items included from U.S. budget (Appendix 6.8) to provide comparability.

[b]Japanese data adjusted by sale weight of 606.9 kg and an exchange rate of ¥227 in 1980 and 629.2 kg and an exchange rate of ¥238 in 1983. United States data simply converted to kg.

Budgets for United States feedlots, detailed in Appendix 6.8, include three types; farmer, commercial and all sizes. A few of the noncash, i.e. economic items are excluded from the summary in Table 6.2 to provide comparability with the Japanese budgets. The results indicate that Japanese costs for feeding Wagyu steers are three times higher than U.S. costs in farmer (i.e. small) feedlots regardless of the cost category; variable, total cash or total economic (Table 6.3).

The major expense in both Japanese and U.S. feeding is the feeder calf, which comprises about 60 percent of expenses in either country (Table 6.3). But, it is the absolute size of the difference which is quite important, for the feeder animal costs three times as much in Japan as the United States. Feed cost is about 4 times as much in Japan as the U.S. mainly because the conversion rate in feeding Wagyu cattle is so low. Consequently, analysis in Table 6.3 shows that when expenses are considered without the cost of feeder calves, all three categories of expenses are still about three times as high for feeding Wagyu cattle as expenses in United States farmer feedlots. Japan's total cost would be 1.95 times that of the United States if feeder calves in Japan cost the same as in the United States.

DAIRY STEER FEEDING CHARACTERISTICS

Dairy cattle feeders are much more concerned about minimizing cost per kg of gain than are Wagyu feeders. Statistics on grading show that the dairy feeder's strategy does not carry an especially high penalty, for about the same percentage of their steers grade Medium as do Wagyu:

	Range, 1980-1983	
Grade	Wagyu	Dairy
Supreme and Superior	6-7	0-0
Excellent	18-26	0.4-1.0
Medium	51-53	52-53
Common	15-24	42-46
Utility	0.9-1.0	3-4

The very different approaches taken to feeding Wagyu and dairy cattle is highlighted by comparing the

Table 6.3. Analysis of Japanese Wagyu feedlot costs versus United States farmer feedlots, 1983

Item	Kyushu Japan	United States	Difference	Ratio, Japan to United States
	-- Dollars per kg sold --			
Expenses with feeder cattle				
Feed	1.61	0.40	1.21	4.03
Nonfeed	0.15	0.11	0.04	1.36
Feeder cattle	2.31	0.77	1.54	3.00
Subtotal	4.07	1.28	2.79	3.18
Manure credit	0.18	0.00	0.18	—
Total variable	3.89	1.28	2.61	3.04
Total cash	4.05	1.39	2.66	2.91
Total economic	4.54	1.52	3.02	2.99
Expenses without feeder cattle				
Total variable	1.58	0.51	1.07	3.10
Total cash	1.74	0.62	1.12	2.81
Total economic	2.23	0.75	1.48	2.97
Expenses with U.S. feeder cattle cost[a]				
Total variable	2.35	1.28	1.07	1.84
Total cash	2.51	1.39	1.12	1.81
Total economic	3.00	1.54	1.46	1.95

Source: Derived from Table 6.2

[a]Assumes that the Japanese cost is $0.77 per kg rather than $2.31.

fattening periods. Whereas Wagyu are on feed for about 17-20 months (essentially from weaning to slaughter) and are sold at about 28-29 months of age, dairy steers and heifers are only on feed for about 13-14 months and are sold at 20-21 months of age (Table 6.4).

Another way to compare feeding practices is to analyze weight gain. Budgets used in this chapter show that Wagyu cattle entered the feeding phase at about 294 kg and were slaughtered at 629 kg. The respective weights for dairy cattle in 1983 were 272 kg and 673 kg. Total gain was thus 335 kg for Wagyu and 402 kg for dairy cattle. But, average daily gain was only 0.64 kg for the 522 days Wagyu were on feed versus 1.06 kg for the 381 days the dairy steers were in the feedlot.

DAIRY STEER FEEDING COSTS AND RETURNS

Budgets for feeding dairy steers to slaughter weight in the Midwestern United States and Hokkaido, Japan during 1980 and 1983 are presented in this section. Results for 1980 are provided in Table 6.5, and for 1983 in Table 6.6. Supporting data are detailed in Appendixes 6.9-6.11. The typical Hokkaido operation presented here has a one-time capacity of 50 head, is family managed and owned, and provides all the labor. The feedlot is full virtually all the time. The feedlot itself is a total confinement facility with concrete floor and gates to partition the cattle into 4 groups. Sawdust bedding, which is replaced every 2 weeks, is utilized during the entire period. Old bedding is sold to adjacent farmers.

The United States system used for comparative purposes is a 300 head capacity lot in the upper Midwest which is also family owned and managed. This operation is not meant to be representative of cattle feeding in the United States, but rather of a comparable operation. As in the Japanese operation all labor is from the family. Similar to the Hokkaido enterprise, the Midwest family farmer also feeds Holsteins, but calves are placed on feed at a much lighter weight, 48 kg versus 272 kg in Japan, and also sold at a much lighter weight, 478 kg compared with 647 kg in Japan. The U.S. lot has a death loss of 1 percent (same as Japan) which means that 297 head are sold after the 361 day feeding period. Average daily gain is 1.19 kg. The U.S. feeder in this actual example, in contrast to many upper Midwest feeders which use large quantities of silage in the Holstein fattening process, uses virtually all concentrates and obtains a kilo of gain for 5.6 kilos of feed. The lot is a slatted floor facility

Table 6.4. Age at slaughter weight and fattening period of Wagyu and dairy cattle, Japan, 1974-1983

Year	Age at slaughter weight		Fattening period	
	Wagyu	Dairy	Wagyu	Dairy
	- - - - - - - - - - - Months - - - - - - - - - - -			
1974	24.8	--	15.3	--
1975	26.9	--	17.5	--
1976	25.3	19.8	15.7	11.3
1977	26.3	20.4	17.0	12.3
1978	27.6	20.8	18.0	12.4
1979	27.6	20.5	17.5	12.3
1980	27.9	20.5	18.1	13.0
1981	28.5	21.3	18.9	13.6
1982	29.3	20.9	19.7	13.3
1983	28.8	20.6	19.2	13.1

Source: The original source is MAFF. Production Cost of Livestock. Tokyo, various issues.

in which the manure collects in sumps and is periodically pumped out and spread on the owner's farmland.

Total investment for 1983 excluding cattle is $1,892 per head in the Hokkaido operation versus $506 in the Midwest enterprise largely because of greater land and building cost in Japan (Table 6.6). On the production cost side, the relatively heavy 272 kg calves in Japan are purchased at $2.77 per kg versus $1.98 for the 48.18 kg calves in the upper Midwest. Feed costs are nearly double per head because feed itself is more expensive and, to a lesser extent, because the Japanese cattle are brought in at heavier weights and sold at heavier weights, thus reducing efficiency in conversion. Hokkaido farmers receive a by-product credit on production costs from the sale of bedding ($36.06 per head) while the Midwest farmer is credited with $10.83 per head sold from the imputed value of manure spread on his fields. The net result is a net basic production cost of $1,669 per head sold (including cattle) in Japan versus $550 per head sold in the Midwestern case.

Noncash costs such as interest or opportunity cost on operating capital, equipment, and buildings, along with depreciation and a charge imputed for labor add another $385 per head for the Japanese operation, about twice that of the $162 for the American farmer.

The cost of gain in Japan is $2.55 per kg when only net basic production costs are taken into account, and $3.73 when all costs are considered. In contrast, the cost of gain for the case study upper Midwestern farmer using these 1983 costs would be $1.14 for the basic production costs only, and $1.54 when all costs are considered. The breakeven prices, i.e. cost of production for the basic production only, and all costs, is $2.63 and $3.23 in the Hokkaido case versus $1.24 and $1.60 in the Midwest. The net result is production cost is twice as high for beef from dairy steers in Japan as in the United States.

Analysis of dairy type cattle feeding operations in France and West Germany indicates that if Japanese dairy producers were to feed whole corn silage during the early and mid-feeding periods, and were to feed cattle as bulls rather than steers, and the cattle were slaughtered at about 500 kg (15-16 months of age) rather than 625-650 kg as is now the practice, production costs could be reduced by 30-40 percent. But, adoption of these cost reducing practices will not likely be forthcoming in the near future due to what has previously been described as "Essen Kulture."

Table 6.5. Estimated costs and returns for case studies on feeding dairy cattle in Hokkaido, Japan and Upper Midwest, USA, 1980

	Hokkaido, Japan		Upper Midwest, USA	
	Total, 50 steer capacity	Per steer sold[a]	Total, 300 steer capacity	Per steer sold[b]
	- - - - - - - - - - - - -	- - - Dollars	- - - - - - - - - - - - -	
Investment				
Land	23,612	477.01	375	1.35
Buildings and capital investment	61,946	1,251.43	68,960	248.95
Equipment	6,183	124.91	61,729	222.85
Cattle	—	—	—	—
Total	91,741	1,853.35	131,064	473.15
Production costs				
Cattle	41,704	842.51	40,068	144.65
Feedstuffs, nonpasture	41,861	845.68	101,238	365.48
Feedstuffs, pasture	962	19.43	—	—
Bedding	1,577	31.86	—	—
Veterinary services and medicine	787	15.90	5,154	18.61
Maintenance	690	13.94	3,960	14.30
Utilities	516	10.42	4,070	14.69
Other	119	2.40	9,460	34.15
Total less cattle	46,512	939.63	123,882	447.23
Total with cattle	88,216	1,782.14	163,950	591.88
By-product value	1,671	33.76	3,000	10.83
Net production cost				
Total less cattle	44,841	905.87	120,882	436.40
Total with cattle	86,545	1,748.38	160,950	581.05
Capital costs				
Interest on operating capital	5,293	106.93	23,609	85.23
Interest on buildings and capital investment	3,717	75.09	11,034	39.83
Interest on equipment	371	7.49	9,877	35.66
Subtotal	9,381	189.51	44,520	160.72

Continued

Table 6.5. Estimated costs and returns for case studies on feeding dairy cattle in Hokkaido, Japan and Upper Midwest, USA, 1980 (continued)

	Hokkaido, Japan		Upper Midwest, USA	
	Total, 50 steer capacity	Per steer sold[a]	Total, 300 steer capacity	Per steer sold[b]
	Dollars			
Ownership costs				
Depreciation	5,400	109.09	3,448	12.74
Taxes	917	18.53	6,173	22.29
Insurance	—	—	—	—
Subtotal	6,317	127.62	9,621	34.73
Labor	5,091	102.85	6,750	24.37
Management	—	—	—	—
Subtotal	5,091	102.85	6,750	24.37
Total other costs	20,790	419.98	60,891	219.82
Total, net production and other costs less cattle	65,631	1,325.85	184,773	667.05
Total, net production and other costs with cattle	107,335	2,168.36	224,841	677.59
Income	107,499	2,171.70	187,925	677.59
Net income				
Above net production costs	20,954	423.32	26,975	90.82
Above all costs	165	3.34	-36,916	-124.30
Breakeven per kg				
Net production costs	2.88		1.22	
All costs	3.59		1.70	
Cost of gain per kg				
Net production costs	2.53		1.02	
All costs	3.70		1.55	

Source: Japan data converted from Yen in Appendix 6.9 using an exchange rate of ¥238 = $1.00. United States data from Appendix 6.11.

[a]49.5 head sold.

[b]297 head sold.

Table 6.6. Estimated costs and returns for case studies on feeding dairy cattle in Hokkaido, Japan and Upper Midwest, USA, 1983

	Hokkaido, Japan		Upper Midwest, USA	
	Total, 50 steer capacity	Per steer sold[a]	Total, 300 steer capacity	Per steer sold[b]
	Dollars			
Investment				
Land	32,029	647.05	375	1.35
Buildings and capital investment	55,294	1,117.05	79,304	267.02
Equipment	6,316	127.60	70,989	239.02
Total	93,639	1,891.70	150,293	506.04
Basic production costs				
Cattle	37,577	759.13	28,620	96.36
Feedstuffs, nonpasture	41,393	836.22	111,354	374.93
Feedstuffs, pasture	1,441	29.11	—	—
Bedding	2,157	43.58	—	—
Veterinary services and medicine	588	11.88	6,084	20.48
Maintenance	427	8.63	4,673	15.73
Utilities	644	13.01	4,803	16.17
Other	162	3.27	11,163	37.59
Total less cattle	46,812	945.70	138,077	464.90
Total with cattle	84,389	1,704.83	166,697	561.26
By-product value	1,785	36.06	3,000	10.83
Net basic production cost				
Total less cattle	45,027	909.64	135,077	454.07
Total with cattle	82,604	1,668.77	163,697	550.43
Capital costs				
Interest on operating capital	5,063	102.28	18,003	60.62
Interest on buildings and capital investment	3,318	67.03	9,516	32.04
Interest on equipment	379	7.66	8,519	28.68
Subtotal	8,760	176.97	36,038	121.34
Ownership costs				
Depreciation	4,957	100.14	4,675	15.72
Taxes	—	—	—	—
Insurance	—	—	—	—
Subtotal	4,957	100.14	4,675	15.72
Labor	5,319	107.45	7,425	25.00
Management	—	—	—	—
Subtotal	5,319	107.45	7,425	25.00
Total other costs	19,036	385.36	48,138	162.06
Total, net basic production costs less cattle	64,063	1,295.00	183,215	616.13
Total, net basic production costs with cattle	101,640	2,054.13	211,835	712.49
Income	97,808	2,017.42	188,723	635.43
Net Income				
Above net basic production costs	15,204	348.65	5,508	19.30
Above all costs	-3,832	-36.71	23,112	77.06
Breakeven per kg				
Net basic production costs	2.63		1.24	
All costs	3.23		1.60	
Cost of gain per kg				
Net basic production costs	2.55		1.14	
All costs	3.73		1.54	

Source: Japan data converted from Yen in Appendix 6.9 using an exchange rate of ¥238 = $1.00. United States data from Appendix 6.11.

[a]49.5 head sold.

[b]297 head sold.

SUMMARY AND CLOSING REMARKS

Beef cattle feeders in Japan are dominated by an all-encompassing goal of having their cattle marble to the maximum possible extent. The tactic which they use in trying for this elusive market is slow feeding. But, data show that less than 7 percent of all Wagyu cattle graded made the top two grades, Supreme and Superior, during the 1980-1983 period. In general, the strategy only results in a low meat to fat ratio, an excessive amount of time on feed, and a very high cost per kilo produced. Dairy cattle feeders, on the other hand, are much more pragmatic. Their average daily gains are much higher than Wagyu feeders (1.06 kg versus 0.64 in 1983). Furthermore, both classes of cattle have about the same proportion grading Medium (51-53 percent for Wagyu steers versus 52-53 percent for dairy steers). Also 15-25 percent of Wagyu steers fall in the lowest two grades (those below Medium) versus 45-50 percent of dairy steers graded.

Cost and returns budgets developed in this chapter reveal that variable costs for Wagyu cow/calf operations in Kyushu, Japan (representative of the country) are about 4 times higher than average cow/calf operations in the Great Plains area of the United States (costs are similar throughout the United States except for the Southeast which has much higher costs). Total costs including economic costs are about 4.8 times higher in Japan. There are a number of reasons for the differences, but size is the primary one.

A second set of budgets for feeding Wagyu steers in Japan to slaughter weight in 1983 were compared with three types of U.S. feeding situations; farmer feedlots, commercial feedlots and an average for all sizes. Comparison with the farmer feedlots shows that Japanese costs are about 2.95 times higher than U.S. costs regardless of the cost category examined; variable, total cash or total economic. The major expense in both countries is the feeder calf which comprises about 60 percent of expenses. But, it is the absolute size of the differences which is important as the feeder animal costs about three times as much to Japanese feedlot operators as does a feeder steer to American operators. If the feeder calf were to cost the same to a Japanese feedlot operator as to an American operator, total variable expenses per kilo of animal sold in Japan would be 1.84 times as much as the American operation.

A third set of budgets are a comparison between Japan and the United States in feeding dairy steers to slaughter weights. The analysis reveals that basic production costs (similar to variable costs) are $2.63 for the 50 head Hokkaido example operation versus $1.24

for an upper Midwest United States operator feeding 300 head. The total (economic) cost for these two operations feeding Holstein steers is $1.60 in the U.S. and $3.23 in Hokkaido. In effect, the Hokkaido cost per kilo sold is 2.12 times as much as the U.S. operator's costs when only basic production costs are considered, and 2.02 times as much when all costs are taken into account.

Overall, the budgets demonstrate that the high cost part of Japan's cattle industry is production of Wagyu calves. The second high cost subsector is feeding Wagyu cattle to slaughter weights. The most efficient operation, and the one with the smallest cost differences from U.S. operations, is feeding dairy cattle to slaughter weights. Japan's cattle industry has undergone remarkable structural changes, particularly in the past decade. Today there are many relatively large feedlots, and relatively large beef breed cow/calf operations are increasingly found. Consequently, the beef breed budgets, which have been developed from highly aggregated regional data tend to overestimate production costs as there are considerable economies available when moving from small, to medium and large size operations. The dairy feedlot budgets also overstate production costs in Japan since dairy cattle are generally fed in large (500 head or more) feedlots rather than the 50 head example provided.

The structural changes to larger scale operations, and thus to lower costs, can be expected to continue in Japan. Part of this process will be adoption of technology now known and also new technology--which is being developed at an almost bewildering pace. A discussion of these technologies, and scenarios about the potential impact on productivity over the next 25 years, is the next chapter's subject.

Part III

Technological Change in Japan: Scenarios to 2010

7
Technological Change: Today and the Future

The past six chapters have been vibrant with the changes taking place in Japan's agricultural, livestock and industrial economies. Farm size is increasing rapidly and becoming ever more commercialized. The cattle industry, long a dormant subsector, has not only awakened but is making great strides forward. Production of beef per head of cattle inventory is among the highest in the world and the ratio promises to continue expanding now that commercial, large scale feedlots have become commonplace. Furthermore, the marketing system for cattle and beef uses some of the most sophisticated procedures in the world such as electronic sales of boxed beef. But, despite all these advances, considerable leeway exists to improve existing operating procedures, as well as to introduce new technologies.

Worldwide, technological changes related to livestock and meat production are occuring so rapidly that even scientists closely associated with each field of endeavor have a difficult time to assess the implications for commercial application. The objective of this chapter is to provide an understanding of the range and breadth of new technologies which have recently been developed for animal agriculture, or can be expected to have commercial application in Japan during the next few decades. It complements the review of crop technologies presented in Chapter 4. The major factors involved are labor, management, cost of the feeder calf, and other growth enhancing technologies. Many of the techniques discussed are now beginning to be applied; others are still in their most basic stages. Together they represent a fascinating array of fruits from modern science.

TECHNOLOGICAL CHANGE AND PRODUCTIVITY

Technological change in Japanese agriculture is now increasing at an increasing rate as Japan is in what Lu, Cline and Quance (1979) have termed the era of "science power." Their research, which was for the United States, shows this new era follows the previous rather short one of mechanical power and a slightly longer one called horse power (Figure 7.1). If cattle power were substituted for horse power, and the dates changed by a few decades, the diagram holds for Japan.

Technology, a human-made resource which can be increased through research and development, has begun to be recognized as a resource because of the complementary relationship among various technologies. This means that imagination, time lag in adoption, and institutional constraints are the major limits to growth in livestock productivity, rather than the traditional three: land, labor and capital. In effect, whereas previous technologies such as mechanization of agriculture were heavily capital intensive, the new ones are more knowledge intensive--a new dimension which is especially uncomfortable to economists since coefficients are not generally available for projections.

Productivity is a measure of the technological efficiency with which resources are converted to commodities and services. Productivity can be measured by ratios of output to a single input either by having both the numerator and denominator measured in physical units, or by constant money values. Total factor productivity presumably is most useful although difficulties abound due to disparate quantities of inputs. Two means used for computing total factor productivity are through index numbers, and production functions. The latter is most useful for the theoretical approach in this section.

A production function describes the physical relationship between a firm's input of resources and its output per unit of time under a given state of technology. As technological change takes place the production function coefficients also change. Because the changes are not continuous, smooth, nor necessarily neutral, it is difficult to specify the functional form. However, the form can be conceptualized by considering a more abstract concept of input mix on the horizontal axis, rather than the usual single input being varied with the rest held constant as is typically depicted in a production function.

The base production function for livestock, shown in Figure 7.2, is shown as P_0 using input mix x_0 yielding output Y_0 by producing at point a. Another alter-

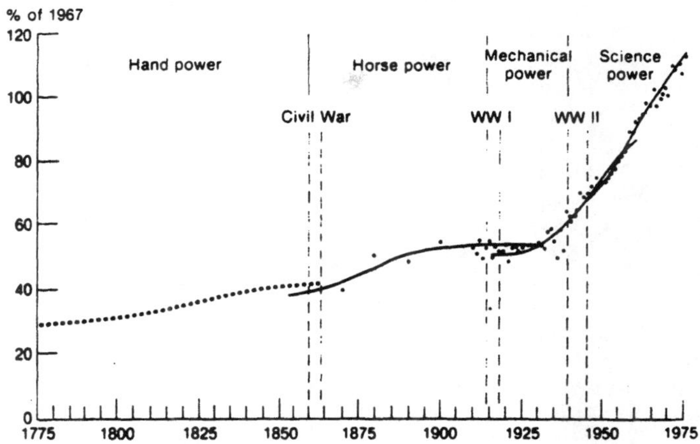

Figure 7.1 U.S. Agricultural productivity growth

Source: Lu, Yao-chi and Leroy Quance. Agricultural Productivity: Expanding the Limits, USDA-AIB 431. Washington, D.C. 1979.

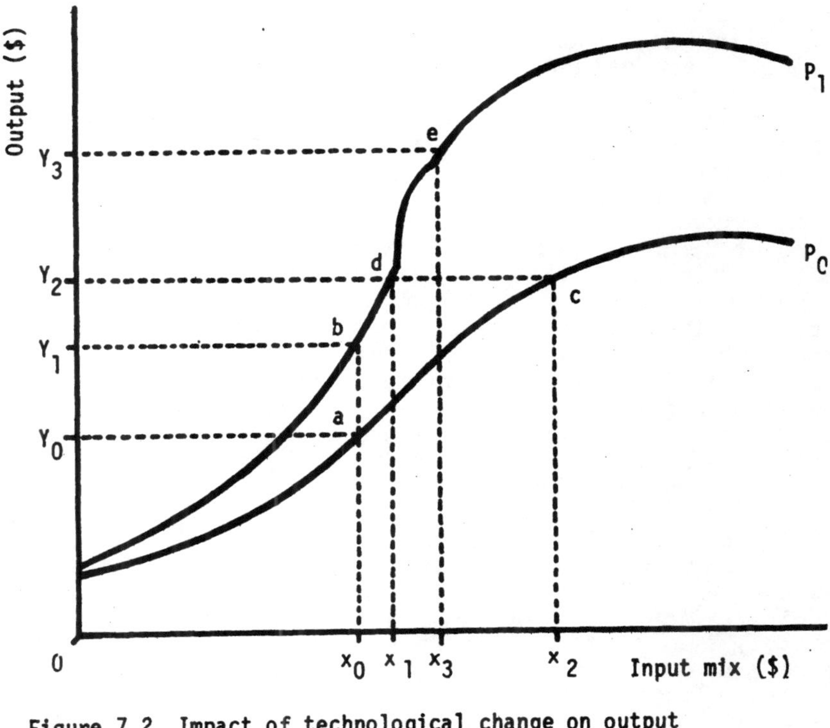

Figure 7.2 Impact of technological change on output

native is use of x_2 input to produce Y_2 output. The impact of technological change can be considered in its simplest form as a movement from point a to point b on a new production function, P_1, at some later time by changing the mix of inputs. As an example, output Y_1 could result from purchase and use of a home computer tied to an individual cow ration distribution system wherein the discounted capital and maintenance cost is offset by a reduction in feed cost. An even higher output, Y_2 can be obtained by increasing the input mix. If, in addition to the home computer and feed distribution system, heat (estrus cycle) detectors were tied to the computer (X_3) an output of Y_3 at point e might be obtained. The important factor here is knowledge and careful manipulation of feedstuffs to optimize feed use.

A crucial point to understand livestock industry technological change during the 1980's and 1990's is to recognize that new production functions will not be smooth, but rather will have "blips," as represented in Figure 7.2 by segment d to e. In this case a rather large output increase from Y_2 to Y_3 can take place at some point in the future with very little additional total input use, again because of expanded knowledge. As an example, before 1990 some dairy farmers in Japan will learn embryo transfer techniques, implant cows with twin male beef breed embryos, grow the calves to feedlot entry weight using a multitude of technologies such as anabolics and feed enhancers, and determine optimum feed use by a supplement optimization computer program. They might then feed the animals to slaughter weight through a cooperatively owned feedlot. More than twice as much beef could be produced per cow under this system.

The last example is somewhat extreme and, clearly, costs and adoption rates, i.e. management, have to be taken into account. But the example serves to point out the importance which knowledge and information management will play in the next few decades in livestock production. Furthermore, the rate of adoption can be expected to increase at an increasing rate as farm size increases and younger farmers, who are trained and psychologically motivated for change, take over both management and ownership of cattle operations.

Feed is widely thought to be a limiting factor to more efficient beef production. But, as was shown in Chapter 4, it can be shipped to Japan very inexpensively. Thus, a trade-off exists--import the final product, i.e. meat, or import feed. From Japan's viewpoint productivity (a physical concept) can be increased regardless whether the feed is imported or produced domestically. Furthermore, efficiency (a monetary concept) can be improved by reorganizing feed use and adopting new technologies and techniques not only related to feed but to all other production processes as well.

TECHNOLOGIES FOR BEEF BREED COW/CALF OPERATIONS

There are numerous research projects underway in both Japan and other countries that can make substantial contribution to improving productivity, and reducing cost in Japan's beef producing operations. The purpose of this and the next three sections is to highlight technologies now available, and ones which will be available during the next twenty-five years. Estimates of the probable impact are given in the next chapter.

One major means to improve beef breed cow/calf operations is by reducing the calving interval, i.e. expanding the calf crop. At present the average calving interval is about 14 months. If this interval were cut to 12 months (i.e. a 100 percent calf crop) production cost would be reduced about 13 percent. There are several means to move toward this goal such as better record keeping on time and duration of heat (estrus) and on aspects related to parturition.

Calf crops are reduced by artificial insemination (compared with natural service) due to the difficulty of locating cows in heat and then inseminating them at the optional time. Heat detection is a major factor since virtually all cows in Japan are artifically inseminated. One way to overcome the timing problem is by estrus synchronization so that all cows can be bred at nearly the same time. Larger operations (50 or more head) can use a cleanup bull to detect females which did not conceive on the first breeding.

Embryo Transfer

Embryo transfer is another technology which can be expected to play an important role in beef breed calf production. The objectives are for (1) breed improvement, and (2) to expand the quantity of calves. Historically, and at present, the beef breed feeder calf is the highest cost part of Japanese fed beef production. For example, data presented in Chapter 6 show that during the feedlot or final finishing phase the calf accounts for about 57 percent of cash costs compared with feed at 40 percent. On an absolute basis the calf entering the feedlot phase is valued at about $1,455 compared with $375 in the United States, i.e nearly four times as much. For this reason special attention is being given in Japan to ways in which feeder calves can be produced less expensively. One approach to dramatically reduce cost is by calving twins (and in a much more restricted sense triplets) rather than single calves (singletons). Freemartins, usually considered a problem with twins are not a constraint since the calves will be fed-out rather than used for breeding.

Embryo transfer (ET) is one way to produce multiple calves that will undoubtedly be used extensively on a commercial basis in Japan in the not too distant future as the technique is increasingly being emphasized. For example, there are researchers at all experiment stations trained in ET. There are also a limited number of commercial veterinarians trained and practicing the technique.

The basic procedure in embryo transfer involves two major steps--and two cows. The first step is to first breed a donor cow by natural service or artificial insemination (AI). About a week later the fertilized egg, or embryo, is recovered nonsurgically from the donor cow by flushing the uterus via the cervix and immediately transferred to a recipient cow, or frozen in liquid nitrogen for later transfer. The donor cow can produce an embryo every 21 days as long as her reproductive cycle is not interrupted by a nine-and-a-half month pregnancy. Furthermore, the donor cow can be "super-ovulated" or hormonally stimulated to produce multiple eggs simultaneously. Occasionally as many as 35 embryos may be recovered, but usually only 5-10 are obtained.

Many improvements have been made in each of the various steps in embyro transfer. But, a major problem area is still the impossibility to be sure a potential donor will provide a suitable number of transferable embyros at a given time. The difficulty is large variation between individuals in their ovulatory response to treatments. More specifically is the existence of poorly responding animals whatever the treatment used. It may be concluded that while strategies can be developed to improve response from these problem animals, it may be more feasible to attempt genetic selection of animals that respond well to superovulation. This latter method seems particularly well adapted to Japan where ET efforts will undoubtedly be coordinated through the national research stations where an organized program for dam selection could easily incorporate selection for superovulatory response.

One way to reduce cost in large herds is by estrus synchronization of recipient cows using a hormone injected into the cows which induces heat in about 72 hours. This procedure is vital because the cycles of the donor and her recipients must be in exact synchrony if the embryo is to survive. However, synchronization is not a problem if frozen embryos are used, except to reduce cost.

Embryo transfer technology has made quantum leaps in just a few years. For example, in the late 1970's there were probably fewer than 5,000 transfers made annually in the United States and these were done by just a handful of very specialized individuals. Commercial application became feasible by the late 1970's when nonsurgical procedures were developed. The first known commercial nonsurgical transfers of bovine embryos were performed in Alberta, Canada in 1981. Baker and Shea (1985) estimate that more than 50,000 bovine embyros were transferred in North America during 1984. Furthermore, by 1984 the technique had become well enough known that some general practice veterinarians were providing this service.

The first ET work was done in Japan in 1957. From 1970 to 1983 over 50 organizations had carried out ET on 5,398 cows. During 1983 alone there were 1,466 ET cows (Iritani, 1984). There has been considerable variation in pregnancy among stations, with rates varying between 15 and 65 percent. Frozen embryos were transferred to a total of 455 cows in 1983. The MAFF started a relatively large scale 3 year project ($430,000) in 1982 entitled "Promotion of the Practical Application of Embryo Transfer in Cattle." It is expected that about 4,600 bovine embryos will have been transferred in 1984 of which about 1,000 would be frozen embryos. There are about 2.7 million dairy and beef cows in Japan which means there would have been one ET for every 587 cows in 1984. There were about 49 million dairy and beef cows in the United States in 1984 so that, if 50,000 embryos were transferred in the United States in 1984 the ratio would be one per every 980 cows, considerably less than in Japan.

An embryo in the United States was split into quarters and calves successfully born in early 1984. The same procedure has also been carried out in swine. Collection of embryos, and splitting in particular, is a complicated procedure which requires considerable skill although it is fast becoming a well understood procedure. But, the transfer process itself, in which the embryo is placed in the recipient cow, is much easier and mainly requires care and practice. It is analagous to artificial insemination although it has been estimated that only about 10-20 percent of professionals now routinely doing AI would be able to develop the required skill to successfully carry out the procedure. Frozen embryos are now being shipped around the world, and ET is fast becoming well-known in many countries.

There are three procedures to produce multiple calves using embryo transfer. The most direct method is to transfer two embryos per recipient cow. The second possibility, but somewhat more complicated, is to breed the recipient using AI or natural service and transfer an additional embryo about a week later. The third possibility is to split embryos upon receipt from the donor cow and transfer them to the recipient cow. A major drawback at this time is that split embryos cannot be frozen so the transfer procedure must take place at the time of collection, thus currently limiting actual application of the technique.

The splitting technique has developed extremely rapidly since it first came into generalized use in the early 1980's. High quality laboratory equipment still costs about $15,000-$20,000, but complete field equipment, while still somewhat awkward, is now available for about $4,000-$5,000. The system used is a Leitz micromanipulator with a microscope that magnifies the embryo 60 times or greater. The micromanipulator has a remote control arm which holds a micro scalpel made from a sliver of razor blade, and another arm to secure the embryo with a suction-holding pipette.

Embryos are carefully examined to determine those suitable for splitting after settling in a dish of holding medium. The splitting procedure is to carefully open the protective outer covering of the embryo (zona pellucida) to split the embryonic mass within the zona. They are then carefully split and half the embryo returned to the zona pellucida of the parent embryo and the other half placed in the protective covering of an unfertilized egg that has had its nucleic material removed and discarded. The result is identical embyros. Splitting of embryos developed mainly during the early 1980's; by mid 1982 at least two firms were engaged in the activity. Their results (Baker and Shea, 1985) have been positive enough that splitting has been permanently adopted as part of their commercial ET program.

The success rate of transferring two embryos to obtain twin or triple calves (the easiest of the three possibilities described earlier) is about 70 percent. This rate is about the same as for first service AI and, in fact, many researchers have obtained higher average success rates than with AI. Thus, in twinning, if 200 embryos were successfully transferred to 100 recipients, 140 calves would result, i.e. a 140 percent born calf crop (calf crop usually refers to the percentage of calves weaned). The average for Japan as a whole, including both dairy and beef breeds, is about 80 percent. (This rate is the annual average and is the result of more than one breeding service on some females).

Estimates of the potential impact from ET must, of course, be based on numerous assumptions, but the main one is the extent to which the technique is used. Calculations in Table 7.1 show that if 20 percent of the national herd's females were transplanted with the potential for twins, a 70 percent success rate obtained, and those females not pregnant serviced the second time by the usual procedure of artificial insemination, the calf crop born would increase 21 percent. If 60 percent of the females received transfers with the potential for twins then calf crop would increase 63 percent. In effect, on the national level, for every 100 cows, there would be 97 calves born under the 20 percent of females transfer level rather than the present 80 calves, and 130 calves under the 60 percent transfer level. ET is indeed potentially the most powerful technology available to the Japanese cattle industry.

Table 7.1. Potential impact for embryo transfers resulting in twins in Japan.

	Percent of national herd females transferred with twins	
Item	20	60
Initial number of cows	100	100
Twin Transfer process		
Cows serviced	20	60
Twin transfer success rate	70%	70%
Cows with twins	14	42
Calves from transfer	28	84
Usual Single Calving		
Usual procedure only	80	40
Left from ET[a]	6	18
Total	86	58
Single success rate	80%	80%
Cows with single calves	69	46
Calves from single calving	69	46
Totals		
Total calves born	97	130
Usual born calf crop	80	80
Additional calves with ET and usual calving	17	50
Increased calf crop	21%	63%

[a]Assumes that cows are pregnancy checked within first month and those found not pregnant are serviced by artificial insemination. ET signifies embryo transfer.

Japanese can, and will, capitalize on more attributes of ET than just multiple births. For example, beef breed embryos can be transferred to dairy breed cows which means that with proper planning many of the dairy herd calves could be of beef breeds thus enhancing the calf's value. Another breakthrough with somewhat longer term, yet wide potential application, is sexing embryos as this could lead to a greater percentage of male calves, a real advantage in the feedlot since they gain faster than heifers.

Sexing of embryos has been carried out since the late 1960's, but difficult and time consuming techniques, the requirement for trained cytogeneticists and the 30 to 50 percent survival of sexed embyros prevents commercial application at present. However, differentiation of sex by nuclear density of interphase cells using a computerized scanning stage with flash photography of phase images and photometric analysis does have considerable possibility (King, 1984). Sexing is receiving considerable research attention in Japan.

The term cloning generally invokes the image of nuclear transplantation. Today, cloning other than by production of identical twins, is not possible. However, in the future there are three main ways to capitalize on some form of vegetative reproduction of desirable phenotypes; growth and multiplication of embyros in vitro, transplantation of nuclei from adult cells after treatment to restore the genomes to the embryonic zygote state, and production of diploid eggs without being fertilized at all. Any one of these three procedures will lead to the true and economic cloning of mammals (Markert, 1984). But, much more basic research is needed before commercial application will be possible.

Other Technologies for Cow/Calf Operations

Productivity, which is often measured by kilos of beef produced per total number of cattle in the national herd, can be increased by obtaining a higher calf crop as discussed in the previous section about multiple births using embryo transfer. Another means, through increased weight per animal, has a wide variety of options, ranging from improved health to nutrition. One way is through genetics, and a practice that has become common in many countries is by controlled crossbreeding.

The process of choosing certain animals of one generation to be parents of the next generation and of determining the number of progeny they will be allowed to have is known as selection. One way to compliment individual selection is to introduce crossbreeding, a genetic technique to take advantage of heterosis, also known as hybrid vigor. There is considerable difference in the amount of heterosis between traits. There is practically no improvement in conformation from crossbreeding, but growth is reasonably heterotic and reproduction, i.e. weaning rate is quite high. Crossbreeding, a mating system that combines hereditary material from two or more pure breeds, will maximize heterosis.

Crossbreeding is useful to incorporate desired characteristics of two or more breeds in complementary combinations. It has been used to overcome environmental difficulty in European breeds, to obtain hybrid vigor or both. The greatest interest in crossbreeding has been between English or Continental breeds (Bos taurus) and Zebu breeds (Bos indicus). As a result, a number of terminal crosses have resulted in new composite or synthetic breeds such as Santa Gertrudis, Beefmaster and Charbray. There are also a number of terminal crosses between English breeds, and English and Continental breeds.

There is little possibility of introducing crossbreeding systems in Japan which incorporate Zebu breeds except possibly in Okinawa where there is a demand for improved heat tolerant cattle. The major demand in Japan, and thus the reason for crossbreeding, is (1) to improve weaning weights (2) to improve average daily gain in feedlots and (3) to improve the carcass quality score as reflected by greater marbling and reduced exterior fat. There are Wagyu cattle in Japan which excel in the above three desirable characteristics. However, much more rapid advances both in terms of time and results could be obtained by crossbreeding rather than concentrating on straight Wagyu breeding, provided a suitable cross could be found.

One approach to crossbreeding is to emphasize improved milk production since Wagyu have very low milking ability and calves must be creep fed from 2 months of age until weaning, i.e. at about 8 months of age. If the milking ability can be improved to the extent that creep feeding will not be necessary, production cost can be reduced about 2-2.5 percent. One hybrid possibility is Wagyu crossed with Holstein. The problem is the large difference in price of carcasses between beef and dairy breeds in Japan. As a consequence, it is unlikely that this cross is economically feasible as a beef breed. But, there is potential for using Wagyu bulls (or through AI) to freshen Holstein cows where the calf will not be needed as a replacement. A not very practical alternative is to use dairy cows as nurse cows by mixing a small number of them with beef breed cows.

Another means to reduce cost of producing beef calves is to expand the use of grazing, a practice hardly used at present. There is very little possibility of expanded grazing land where cattle are now raised (usually in heavy agricultural areas), but there are substantial, albeit marginal, grazing lands in mountainous areas. There has been some use of grazing lands from spring to fall, but they have not been utilized to the fullest extent due to the poor milking ability of mother cows. Producers who have had access to marginal lands have not used them to any great extent because calf size was too small at weaning time. A further problem is additional management associated with grazing such as expanded health practices like vaccination against piroplasmosis and tick control. Estimates of additional marginal land availability are provided in Chapter 8.

Death loss is another aspect to which attention will be paid. At present, the rate in beef breeds is about 4 percent, mainly from diarrhea and pneumonia which are brought about by improper feeding, dirty facilities, group feeding and introduction of sick animals into the barn.

The improvement of low quality roughages is another major way to reduce beef breed calf production cost. Rice straw is the principal roughage which is now used, being fed at the rate of about 6 kg per day. That level is potentially equivalent to about 58 percent of a cow's TDN requirement, but the digestibility of rice straw can be increased 10 percent by treating it with ammonia. This means a potential reduction of 5.8 percent in TDN requirement. Various studies show that 100 liters of 35 percent ammonia are required for 1 ton of straw with the treatment requiring 3 weeks sealed in an airtight container such as an old barn, shed or silo. This method is economical in Japan and can be expected to serve as the basis for improvement of rice straw during the 1980's.

Another means to improve low quality roughages is by an explosion process. A MAFF report shows, for example, that TDN content of white birch wood chips increased from less than 10 percent to 60 percent by using the explosion technique. Reports on rice straw show that in-vitro digestibility of dry matter increases from 44 percent to 69-77 percent. However, there seems to be a large variation in the effect of the explosion treatment in different kinds of raw materials such as soft and hard woods, and from different experimental procedures such as pressure and time. Furthermore, the palatability of exploded materials is not high due to foul odor and acidity. Overall, it appears that while this technique has considerable potential for the longer term, there will be little or no commercial use made of it before 1995.

Calf production cost can be reduced by shortening the period to weaning or increasing the calf's size. The objective, at least from a national viewpoint, should be to grow an animal with the largest frame possible. One way to produce a large calf which is not over-fat is by using growth stimulants. Zeranol is one which, while not yet approved in Japan, has considerable potential as a cost reducing/weight increasing method. Worldwide studies show that suckling calves implanted with zeranol gain 10-14 percent more in body weight during the treatment period (90 days) with some improvement in feed efficiency. There is no residue problem and no human or animal safety problem. There are now two growth stimulants sold in Japan, one under the brand name Synovex and a generic of that brand.

TECHNOLOGIES FOR BEEF BREED FINISHING OPERATIONS

Price differences between cattle grades in Japan, as discussed in Chapter 5, are remarkably large. As a consequence, there is incentive for many beef breed producers to strive for higher than calculated economically optimal grades by prolonging the fattening period. This does not mean all producers reach for the top grade--they don't. Rather, it simply means virtually all beef breed cattle are fed in a way which carries them beyond their economic optimum. Recently, the fattening period for dairy cattle has also been prolonged to produce beef of relatively good quality, even though they produce lower quality beef than beef breeds. The present typical systems for producing domestic fed-cattle are shown in Table 7.2. There is, of course, considerable variation between operations. For example, the beef cattle example in Chapter 6 is based on an average time on feed of 17 rather than 19.7 months.

Japanese beef breed cattle are not large-framed animals. The adult weight for cows is about 560 kg, and 940 kg for bulls. Therefore, as might be expected, daily gain declines dramatically during long feeding periods when the weight of steers exceeds 500 kg. This is one reason the average daily gain during the entire finishing period is just 0.59 kg for beef breeds. It is not feasible, from the standpoint of biological or economic efficiency, to finish steers to 640 kg, yet that is the average today.

Producers believe a long fattening period is the most profitable approach to finishing cattle because they look at total gross income per head rather than net income per unit of resource per year. Furthermore many, if not most, feeders deliberately slow down daily gain to achieve a higher degree of marbling. But, from an economic perspective, most producers suffer severe opportunity costs.

Table 7.2 Present systems for finishing steers to slaughter weight in Japan.

Breed	Placement on feed Age	Placement on feed Weight	Fattening period
	- Mo -	- Kg -	- - Mo - -
Beef	9.6	285	19.7
Dairy	7.6	270	13.3
Breed	**Slaughter Age**	**Slaughter Weight**	**Average daily gain**
	- Mo -	- - - - Kg	- - - - -
Beef	29.3	640	0.59
Dairy	20.1	655	0.96

Source: Estimates based on several MAFF publications.

As was pointed out in Chapter 5, only 20 percent of beef breed steers graded are Excellent (Jyo) or higher. There are no reliable strains of cattle in Japan that will consistently produce highly marbled beef even when fed for long periods. Virtually all cattle coming out of feedlots are over-fat with the result that a high degree of trimming is required on carcasses.

The Ministry of Agriculture (MAFF) has recognized the overfinishing problem and, as a result, set official guidelines for length on feed and age at slaughter. For dairy steers, slaughter age is targeted to 18 months with average daily gain in the feeding period of 1.1 kg. The slaughter age for beef breed steers is set at 24 months or less and an average daily gain of 0.8 kg. But, even those target fattening periods seem much too long for both breeds to effectively reduce total and average cost. It is clear that optimal length on feed could be even shorter than the MAFF targets, average daily gain higher, and final weight much lower than at present.

A first step to lower production cost would be to recognize the need to reduce genetic variation in quality of beef breeds. That means both a private and government initiative on breed improvement. Since a long time is required to improve breeding that means, for the short term, the need to divide beef breed finishing systems into two parts. One group, which would incorporate the majority of beef breed steers, would target final grade at Medium (Chu). This beef is relatively lean and thus the whole emphasis would be to improve gain as much as possible and, alternatively, to reduce cost as much as possible.

The second part would be heavily marbled cattle, i.e. what has been the "traditional" approach. Some producers have the expertise to meet this market and they should be assisted by concentrating on breed improvement targeted both at gainability and marbling. In effect, there needs to be recognition that the market is differentiated and breeding programs developed to meet each of these market shares. It is through this recognition that a rational plan for technology adoption can be formulated and executed.

Prior to setting forth a long term breeding plan there needs to be discussion about whether targets can be most easily reached by restricting breeding programs to Japanese beef breeds, or whether foreign breeds would be helpful in improving genetic uniformity. However difficult, a decision must be reached for, at present, the system can be characterized as simply a "grasp all--lose all" approach. "Grasp all" means attempting to obtain both high marbling and quantity, while "lose all" means actually failing in the quality target and simply producing beef for popular type uses.

The approach which will increasingly be taken in Japan will be targeting production to an evermore differentiated market demand for beef. For example, whereas beef breeds are now feed for about 19.7 months and slaughtered at 640 kg, it is expected that by 1990 that time will drop to 16 months and slaughter weight to 600 kg (Table 7.3). In effect, targets for amount of marbling will change in recognition of both markets and economics.

Changes in the approach to feeding cattle are illustrated in Figure 7.3. In the case of beef breeds, two systems will emerge, one which might be called "ordinary" type fed cattle while the other is "special" as it aims at doing a good efficient job of reaching the highest grades. As shown, the range between grades in the ordinary system will tighten up resulting in the percentage of ordinary type increasing. It is likely that by 1990 silage based finishing of dairy cattle will become more common with the result that what might be called an "extensive" fed cattle system will develop. Then, by 2000, the range in grade of dairy fedcattle will narrow as concentrate and silage feeding are combined and producers target for the middle grades.

The scenarios developed in Table 7.3 and Figure 7.3 should not be considered projections. In fact, they are not even targets but rather are part of a simulation exercise which continues through the next

Table 7.3. Present and proposed systems for finishing steers to slaughter weight in Japan

Year	Placement on feed: Age	Placement on feed: Weight	Fattening period	Slaughter: Age	Slaughter: Weight	Average daily gain
	- Mo -	- Kg -	- - - Mo	- - -	- - - Kg	- - -
			Beef breeds			
Present	9.6	285	19.7	29.3	640	0.59
1990	8.0	240	16.0	24.0	600	0.74
2000	7.0	250	10.0	17.0	550	1.00
2010	6.0	260	8.0	15.0	550	1.21
			Dairy steer (Intensive)[a]			
Present	7.6	270	13.3	20.1	655	0.96
1990	7.0	250	10.0	17.0	630	1.25
2000	7.0	250	8.0	15.0	570	1.33
2010	7.0	250	8.0	15.0	570	1.33
			Dairy steer (Extensive)[b]			
Present	--	--	--	--	--	--
1990	7.0	250	8.0	15.0	570	1.33
2000	7.0	250	8.0	15.0	570	1.33
2010	7.0	250	8.0	15.0	570	1.33

[a]Total confinement on commercial concentrates.

[b]Large scale operation based on pasture and whole crop silage.

PLATE 1. Wagyu cows have traditionally been raised on very small operations and spend virtually all their lives confined in a barn.

PLATE 2. Modern Wagyu cow/calf operations have 10-20 cows and heifers. They are permitted access to loafing areas even in the upper half of Japan which is covered with snow 4-5 months per year.

PLATE 3. Wagyu calves are carefully groomed prior to entering the auction ring. Virtually all Wagyu calves are sold through auctions.

PLATE 4. Wagyu calves are sold individually at auction. Potential buyers have access to extensive pedigree information even though most calves are sold for feeders.

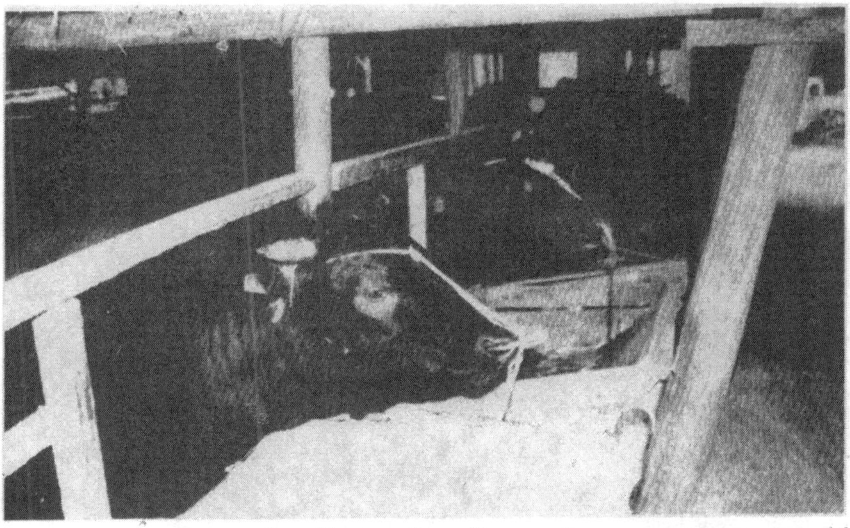

PLATE 5. Modern feedlots for Wagyu have sprung up all over Japan. There is no stocker period on pasture so Wagyu are maintained in a confinement feedlot from weaning to slaughter.

PLATE 6. Crossbreeding to produce feeder calves or for development of a composite beef breed offers great potential to expand productivity. But, adoption of this production technique will likely be slow in coming.

PLATE 7. Dairy cattle are maintained in a confinement feedlot from about four months of age to slaughter. Bedding, changed every 7-10 days, is used in most operations.

PLATE 8. There are a wide variety of feedlot facilities. Producers, especially of dairy steers, are continually experimenting to reduce cost and increase gain.

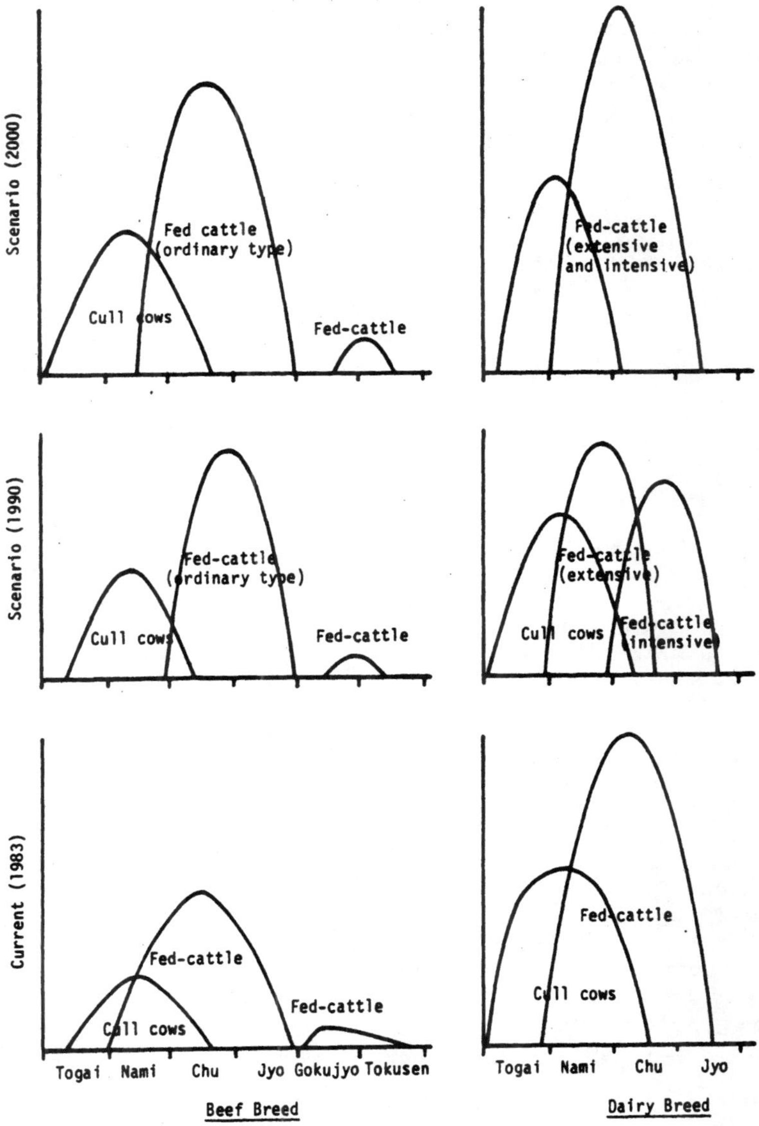

Figure 7.3 Domestic beef supply by carcass grade

two chapters. In effect, it is clear that change will take place and that is should be guided. However, our role, and the spirit of this long-term outlook work, is to evaluate likely situations, ones which are based on needed and plausible structural and technical adjustments. Each of the systems is reasonable and the indicated coefficients will likely be attained.

The fattening period can be shortened to those indicated without seriously inhibiting production of high quality beef. Furthermore, the beef breed coefficients outlined are averages since there are two major subsystems (high quality and popular) involved. However, due to basic economic pressures, and other demand side factors presented in Chapter 2, there will be continued shift from producers targeting for the high quality market to the popular one.

There are several objectives to be attained by reducing production cost. On the supply side it means more potential for expanded domestic production, i.e. greater competitiveness by domestic producers with imported products. On the demand side it means reduced cost to consumers. Shortening the feeding period dramatically, as shown in Table 7.3, will result in a major savings in feed. At present about 1 ton of TDN (total digestible nutrients) is required for each 102 kg of beef (carcass weight basis) from beef breed animals, and 143 kg from dairy breeds. But, if the scenarios developed in Table 7.3 were met, the output in 1990 for each ton of TDN would be 136 kg from beef (a 30 percent increase) and 154 kg from dairy breeds (a 10 percent increase). By the year 2000 the output from beef breeds would be 188 kg per ton of TDN (an 80 percent increase) and 189 kg from dairy breeds (a 30 percent increase).

The fattening period for beef breed cattle can be reduced through breed improvement and better feed efficiency. It has been shown that in the United States use of Zeranol improves gains by up to 10 kg per implant and improves feed efficiency by 8-10 percent in feedlot cattle. Another means to improve daily gain as discussed in the last chapter is by feeding intact bulls. The United Kingdom experience is for animals to go to slaughter at 12 months of age weighing 420 kg. Bull feeding is not practiced in the United States due to the grading system. It is also unlikely to have much place in Japan due to the grading system and tradition.

There are definite economies of size from feeding cattle in large feedlots (500 head or more). Commercial feeds can be obtained at a volume discount of at least 11 percent over 20 kg paper sacks. A reduction of that magnitude means a difference in production cost of about 3 percent for beef breeds and 4.5 percent for dairy breeds since concentrate feed constitutes about

36 percent of total cash cost in a beef fattening operation and 47 percent in a dairy steer finishing operation. These benefits are of such a magnitude that the small scale finisher can be expected to continue to drop out and for average feedlot size to grow rather rapidly. It is not unreasonable to expect that very large scale feedlots (10,000-30,000 head capacity) will be forthcoming in what might be called a beef kombinat (beef complex). The bottleneck will be supply of calves, so it is most likely such a complex would be located in areas with access to large numbers of calves such as Hokkaido, Tohoku and the southeastern part of Kyushu. It is even possible calves will be imported in increased numbers from Australia or China.

Advances and considerable changes will be made in feeds and feedstuffs. For example, use of whole crop silage can be expected to increase as technological advances in crop production are made. There are also a variety of by-products from food processing factories which are not now being utilized to any great extent but which have potential.

Death loss rates increase with longer fattening periods since high concentrate rations contain high levels of readily-fermentable carbohydrates which cause metabolic disorders. This is compounded by the practice of limiting rations to slow down feed intake. At present, feedlot death loss nationwide is about 3.4 percent. That rate could easily fall to less than one percent with radically shortened feeding periods and improved management available in large scale feedlots.

TECHNOLOGIES FOR DAIRY CALF PRODUCTION

Dairy calves for beef, as explained in the last chapter, are essentially a by-product of the dairy industry. As such, they are not well cared for on dairy farms and consequently death loss is high, about 15 percent. One main problem is that many newborn calves are not fed a sufficient supply of colostrum. It is recognized that providing mother's milk will cut death losses considerably. Also, fermented colostrum can be useful as a nutritious supplement.

Another problem in many dairies is unsanitary conditions. Calves should be kept separately in individual hutches to avoid microbial infection. They can also be maintained on a concrete floor if sufficient bedding is provided. In brief, management is the key for newborn calves and growing them out to 3-4 months at which time they go into a finishing operation. Given that 70 percent of Japan's beef production is from dairy breeds, a 5 percent reduction in death loss of dairy calves could potentially result in a 3.5 percent increase in beef production.

TECHNOLOGIES FOR DAIRY STEER PRODUCTION

Most of the previous discussion on beef breeds steers also holds for dairy steers. Because this subsector is already relatively efficient less potential exists for improvement on a percentage basis than beef breeds. But, because dairy beef accounts for such a large percent of beef production, it deserves considerable emphasis.

SUMMARY AND CONCLUSIONS

A popular belief both within and outside Japan is that, because there is such a small amount of arable land, the cost of calves will always remain high. But, apart from shifts in land use taking place, there is also a vast amount of potential new technology--as well as a very large pool of known technology--which will be taken advantage of in the next decade. It now appears that the principal ones which could result in reduced calf cost in Japan are multiple births by embryo transfer (ET) and improved management.

Most use of ET will probably be in the dairy sector where, one day, transfer of sexed, breed specific, twin embryos will largely replace artificial insemination on large dairies. In addition, many dry-lot beef breed cow/calf operations could also be involved. Clearly, ET as a means to produce multiple calves will not be used on all operations or even on the majority of them in the near future. Nevertheless, with new technologies developing in sexing embryos and delivering them without freezing (thus improving fertilization rate), within a decade Japan's total calf crop could potentially be increased from 25 to 35 percent by this technique alone. ET is now known and practiced on every research station in Japan and is beginning to be used commercially. Widespread adoption is just a matter of time as the basic elements are in place. In fact, by the early 1990's, ET will be nearly as common as artificial insemination. The techniques related to it, and adoption of it, are rapidly becoming a people rather than technical problem.

It can be concluded that Japan's beef production industry is now well into what can be called the "takeoff" stage. The world is entering an era in which the major technological advances will come from <u>knowledge</u> and emphasis on doing things well. The paradox is that the new technologies will be widely applicable in Japan simply because (1) their operations are very intensive, (2) product prices are now very high which permits use of increased inputs, (3) seventy percent of the beef is derived from dairy cattle and that is the subsector where much of the new technology is most appropriate and likely to be adopted and (4) much of the new technology can best be developed and adopted via strong national level administration.

The next chapter is devoted to analyzing, from an economic perspective in particular, the benefits and costs from technology adoption. Major emphasis is placed on quantifying the extent to which beef production can be expanded.

8 Impact of Technological Change

The impressive array of new technologies presented in the last chapter points to the potential for a radical decline in Japan's beef production cost. Furthermore, now that problems in production of other livestock and poultry as well as crops have essentially been solved, it is only natural that governmental efforts be redirected to place more emphasis on beef production. Some areas now receiving greater attention include roughage source development, storage, feeding methods, and national breed development and selection programs. The purpose of this chapter is to measure the potential for (1) productivity enhancement (a physical concept), and (2) cost reduction (an efficiency concept) from various technology adoption and industry structure change scenarios.

BEEF AND MILK PRODUCTION: JAPAN IN AN INTERNATIONAL PERSPECTIVE

There are a number of measures which can be used to evaluate cattle industry productivity. An example at the cow/calf level is kilos of weaned calf per cow, while in the feedlot useful measures are average daily gain, feed conversion and grading. The difficulty is appropriate statistics are not readily available. However, one exception is beef production per head of inventory, which can be calculated by dividing production by total cattle inventory. This measure, while somewhat crude, shows that although Japan's cattle industry is struggling against formidable constraints such as a small proportion of arable land, little available pasture land, and historical exigencies, production per head of inventory stood at 109.4 kg in 1984, considerably higher than United States productiv-

Table 8.1. Cattle inventory, beef production and beef production per head of inventory, Japan, EEC-10 and the United States, 1960-1984

Item	Year					
	1960	1965	1970	1975	1980	1984
Cattle inventory (1,000 head)						
Japan	3,163	3,175	3,953	3,644	4,248	4,664
EEC-10	66,705	69,164	73,926	80,551	77,784	79,491
United States	96,236	109,000	112,369	132,028	111,192	114,040
Beef production (1,000 MT)						
Japan	143	216	278	353	418	510
EEC-10	4,450	4,835	6,041	6,723	7,026	7,281
United States	7,195	8,957	10,103	11,271	9,999	10,755
Beef production per head of inventory (kg)						
Japan	45.2	68.0	70.3	96.9	98.4	109.4
EEC-10	66.7	69.9	81.7	83.5	90.3	91.6
United States	74.8	82.2	89.9	85.4	89.9	94.3

Source: Adapted from data in USDA, Livestock and Meat Circular, FLM 10-78, FL&P 1-84, and FL&P 2-84.

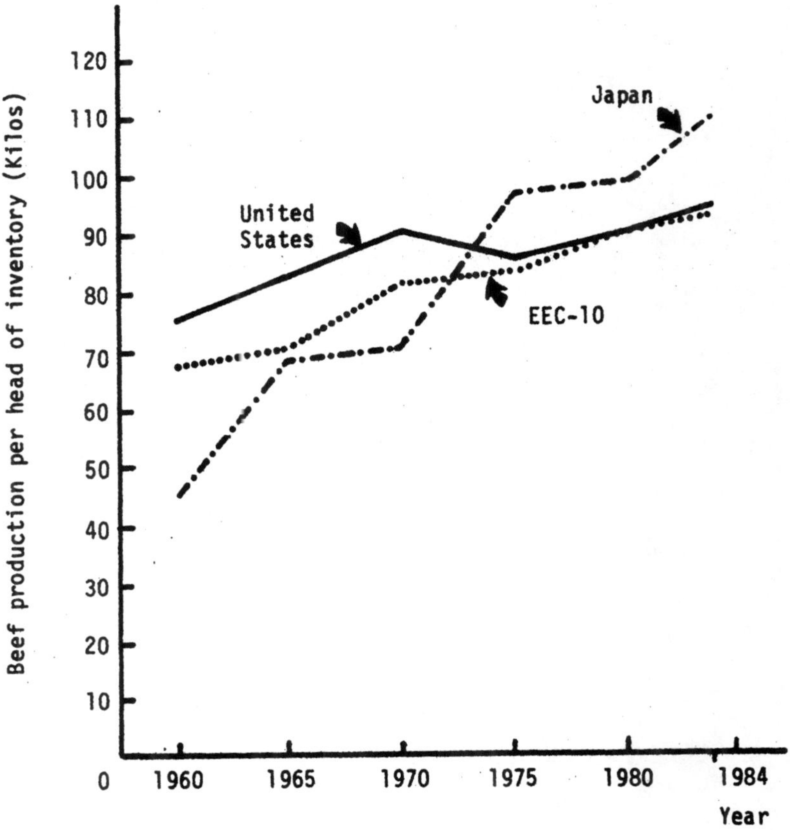

Figure 8.1 Beef production per head of inventory, Japan, EEC-10 and the United States, 1960-1984

ity of 94.3 kg, or the 91.6 kg registered by the EEC-10 (Table 8.1 and Figure 8.1).

Calf crop in Japan is about 80 percent while it has been about 79 percent in the United States. Japan's calf crop is relatively low given the intensity of operations due to use of artificial insemination rather than natural mating. But, care must be taken to properly interpret production per head of inventory. For example, one main reason the percentage in Japan is high is because cattle are sold out of feedlots at heavy weights, about 640 kg versus 450 kg in the United States.

The EEC-10 has a computed beef production per head of inventory about equal to the United States. The steady improvement in EEC productivity of about 1.3 percent annually from 1960 to 1984 is principally due to expanded use of silage resulting from improved short season maize varieties. Furthermore, techniques of feeding cattle to slaughter weight have improved and a greater proportion of calves are continuously being fed to heavy weights rather than being slaughtered as vealers.

Japan's beef productivity has increased steadily at a rate of 3.8 percent annually since 1960, when the average was just 42.5 kg per head of inventory. One major factor is that in the early years virtually all calves other than breeding stock were kept as replacements or were slaughtered as calves, largely for sausage. Furthermore, the average level of management has improved rapidly, a characteristic associated with the decline in number of farms with beef cattle. Finally, feedlot industry development, along with national infrastructure such as improved roads and coordination between regions, now permit smooth flow of livestock and products over relatively large distances. Overall, despite drawbacks in this measure, it does show that Japan has a much higher productivity than is often believed.

A second measure of productivity which is very important is milk yield since beef and milk are joint products of the dairy industry. Japan has made great strides in milk production per cow in lactation and now surpasses the EEC-10 by about 15 percent. Japan's greater productivity is largely due to their industry being based almost exclusively on Holsteins, a pure dairy breed, while EEC milk is mainly produced by dual purpose cattle. The United States average in 1982 was 5,637 kg per cow, which is about 22 percent more than the 4,633 kg recorded in Japan (Table 8.2 and Figure 8.2). The United States dairy industry is founded exclusively on dairy breeds, mainly Holstein which, when combined with a continual growth in farm size, has led to the spectacular advances in production per cow.

PRODUCTIVITY INCREASE ESTIMATES

The purpose of this section is to provide estimates of the impact from technology on the four main beef production systems; beef breed calves, dairy breed calves, and beef breed cattle and dairy cattle in feedlots. The estimates for each system are then amalgamated into estimates of beef production per head of inventory out to the year 2010. The economic impact is then quantified. The first step to accomplish this section's purpose of calculating productivity increase for the next 25 years is to estimate the impact on output of four major areas; breeding, reproduction, nutrition, and management.

Beef Breed Calves

The estimates for beef breed calves, as for all systems, are a subjective weighted average of technical potential and likelihood of adoption. For example, sire evaluation programs could <u>technically</u> bring about a 5-10 percent increase in weaned calf weight within 15 years <u>if</u> the programs were initiated immediately on all sires. But, the reality is expanded use of sire evaluation programs will take years, and there is a lag time for benefits to be achieved due to biological reasons. Consequently, our estimate is output of beef breed calves (as measured by increased weaned calf weight) will increase 0-1 percent by 1990, 1-2 percent by 2000 and 2-4 percent by 2010 as a result of sire evaluation programs (Appendix 8.1). A second variable in the breeding area is crossbreeding. A good national level program could probably result in weaning weight increases of 10-15 percent, but we only estimate a 1-2 percent increase even out in 2010 simply due to the problem of overcoming inertia of customs discriminating against this technology.

Estimates of productivity increases for each of the other three areas, i.e. reproduction, nutrition and management are provided in Appendix 8.1 according to the type of benefit realized. The next step, and one also included in Appendix 8.1., is to place the estimates on the basis of additional kilos of calf produced per 100 cows, a procedure that is necessary due to differing benefits. For example, the benefit from increasing calving percent is different than the benefit from increased weaning weight. The various conversion factors provided in Appendix 8.5 are used to obtain the additional kilos produced. As an example, the estimates of 0-1 percent improvement in 1990 for the beef breed sire evaluation programs translate to an additional 0-221 kg per 100 cows.

Table 8.2. Dairy cows in lactation, milk production, and milk production per cow in lactation, Japan, EEC-10 and the United States, 1961-1965 to 1982

Item	Year		
	1961-65	1969-71	1982
Dairy cows in lactation (1,000 head)			
Japan	647	1,180	1,457
EEC-10	28,423	28,024	28,659
United States	16,195	12,049	10,919
Milk production (1,000 MT)			
Japan	2,713	4,697	6,750
EEC-10	86,732	92,617	115,274
United States	56,999	53,173	61,553
Milk production per cow in lactation (Kg)			
Japan	4,193	3,981	4,633
EEC-10	3,051	3,305	4,022
United States	3,519	4,413	5,637

Source: FAO, _Production Yearbook_, 1976, 1981 and 1982.

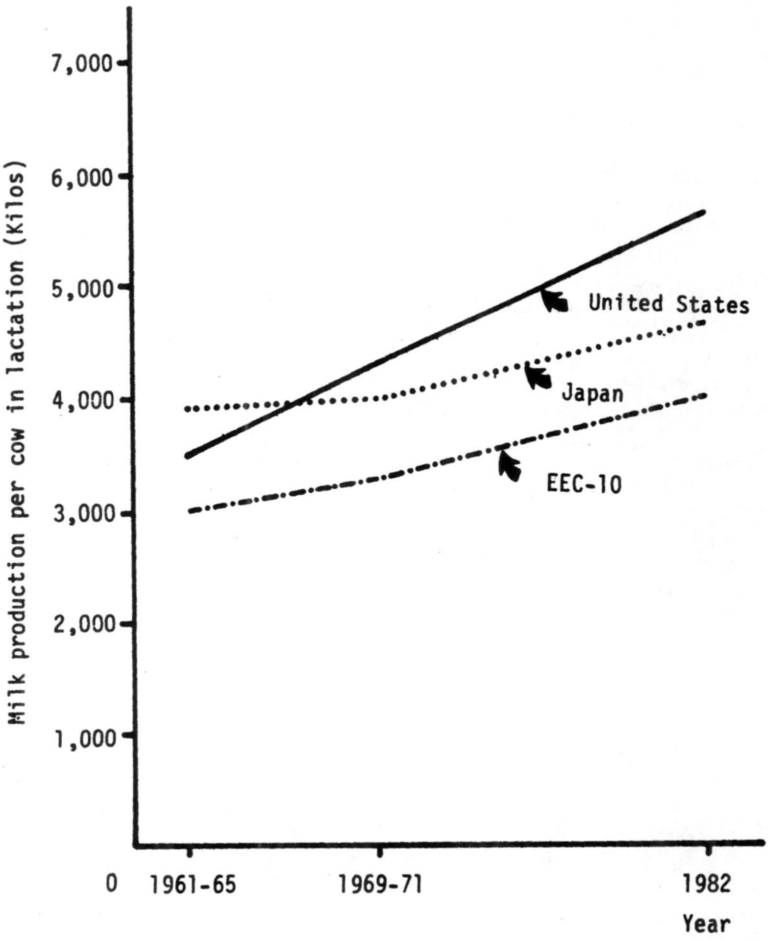

Figure 8.2 Milk production per cow in lactation, Japan, EEC-10 and the United States, 1961-65 to 1982

The additional kilos of beef produced are summed for each of the 3 scenario years. Then, the totals are divided by the total weight of calves per 100 cows in the base year of 1984 and percent increases obtained. The ranges of additional kilos of beef obtained are fairly large--4-15 percent in 1990, 17-39 percent in 2000 and between 30 and 55 percent in 2010 (Appendix 8.1). The averages are 10, 28 and 33 percent for the three periods respectively. These percentages can be interpreted as increases in kilos of beef produced per beef breed cow in the national inventory.

Dairy Breed Calves

Productivity increases for dairy calves turn out to be less than for the beef breed cow/calf system since there is no benefit estimated from breeding advances. There is also less potential from management since only the calf is involved, given that calves are removed from the dam almost immediately after birth.

The dairy calf productivity increases, just as with beef breed calves, are converted to kilos per 100 cows. However, output is defined as calves entering feedlots. The conversion factors in Appendix 8.5 are employed for dairy calves as well as beef breed calves. The productivity estimates are for 1-9 percent more kilos of dairy calf produced per dairy cow in 1990, 13-31 percent more kg in 2000, and 20-39 percent more in 2010. The averages are 5, 22 and 30 percent. Care must be taken to avoid misinterpretation of these as well as the beef calf estimates. They are percentage increases per cow and not per calf.

Beef Breed Feedlots

The impact of tehnological changes and techniques on finishing cattle in beef breed feedlots is broken down into the same categories as for calves; breeding, reproduction, nutrition, and management (Appendix 8.3). Those categories are then separated into 3 groups; growth rate, feed efficiency, and other economic efficiency. The items contained in growth rate are physical in nature and thus constitute the factors for productivity enhancement or, as defined in this chapter, additional kilos of beef produced per head of national cattle inventory. The growth rate factors, which result in mean productivity increases of 5, 13, and 18 percent for the three scenario years of 1990, 2000, and 2010 have ranges of 2-8 percent, 8-18 percent and 12-24 percent.

Feed efficiency enhancement primarily results in reduced cost per animal fed rather than expanded output per head of cattle in the national inventory. The main benefit is from cattle being on feed a shorter period of time and that is an economic benefit. Furthermore, as discussed earlier, a shorter period of time on feed will result in productivity per head of inventory actually declining. Some of the feed efficiency factors have been designated in the category called growth rate, i.e., there is some arbitrariness about category designation. But, this category designation decision turns out to have virtually no impact on the final results. Finally, since feed efficiency manifests itself as an economic factor it, along with the group called "other economic efficiency," are not included in productivity enhancement estimates, but rather are used in the economic analyses.

Dairy Breed Feedlots

The same approach to estimate productivity increases in dairy breed feedlots is used as with beef breeds. However, productivity increases are considerably less, with means of just 2, 6, and 8 percent for 1990, 2000 and 2010 (Appendix 8.4). Estimates for the dairy breed are lower than for beef breeds since dairy feedlots are already operating relatively efficiently.

Summary of Productivity Scenarios

The estimates of productivity increases for the four systems (two calf and two feedlot) from Appendixes 8.1-8.4 are transformed into kilos of beef per head of inventory in Appendix 8.6. A summary of results, including productivity changes for each of the four systems, is provided in Table 8.3. The scenarios indicate that the greatest percentage increases can be expected in beef breed calf production followed by dairy calves. The largest advances in feeding cattle to slaughter weights have already been achieved just by introduction of commercial feedlots.

Production per head of all cattle inventory is not expected to increase nearly as fast as the period 1960-1984, even under the high scenario estimates in which it is calculated to reach 115.09 kg in 1990, 131.1 kg in 2000 and 140.0 kg in 2010 (Table 8.3 and Figure 8.3). The low estimate scenario is productivity per head of inventory at 110.1 kg in 1990, 120.3 kg in 2000 and 123.1 kg in 2010. In contrast, the productivity was 109.4 kg in 1984. The annual rate of growth was 3.7 percent from 1960 through 1984. The low estimate scenario is a growth rate of 0.5 percent annually from 1984 through 2010, while the high estimate for the 26

Table 8.3. Scenarios of productivity increases and beef production per head of cattle inventory in Japan, 1990, 2000 and 2010

Year	Mean	Low	High
	- - - - - - - - - Percent - - - - - - - - -		
	Beef calf productivity changes		
1990	10	04	15
2000	28	17	39
2010	43	30	55
	Dairy calf productivity changes		
1990	05	01	09
2000	22	13	31
2010	30	20	39
	Beef breed feedlot productivity changes		
1990	05	02	08
2000	13	08	18
2010	18	12	24
	Dairy breed feedlot productivity changes		
1990	02	00	04
2000	06	03	09
2010	08	05	11
	Production per head of inventory		
	- - - - - - - - - - kg - - - - - - - - -		
1990	122.8	110.1	115.9
2000	125.5	120.3	131.1
2010	131.3	123.1	140.0

Source: Appendix 8.6.

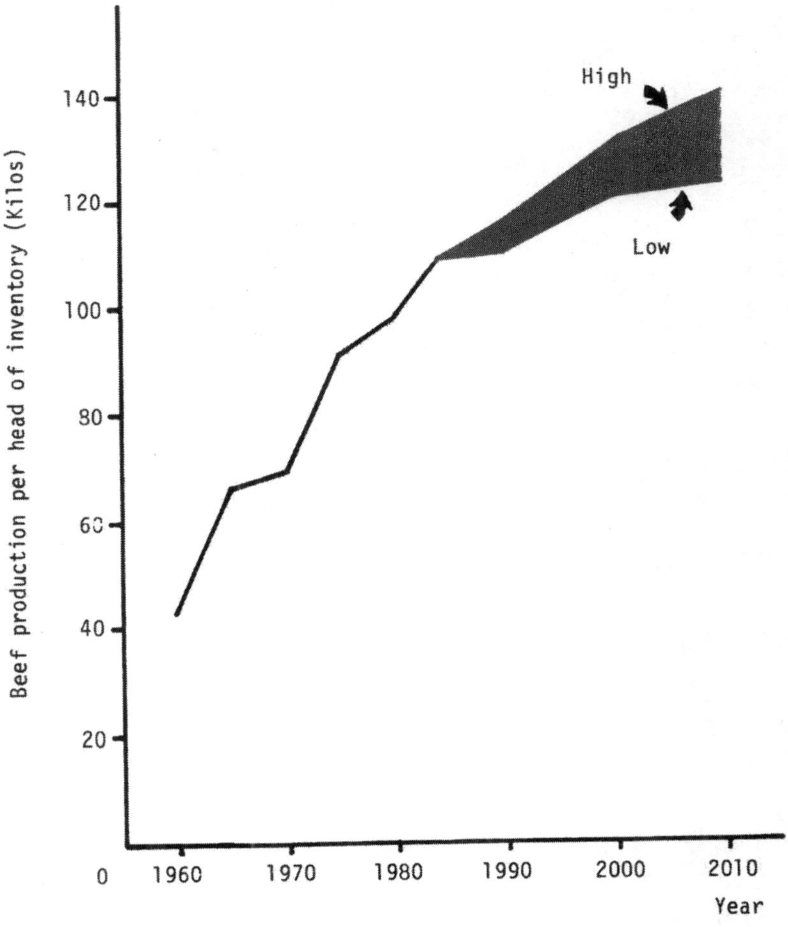

Figure 8.3 Beef production per head of inventory in Japan 1960-1984 with scenarios for 1990, 2000 and 2010

year period is 1.0 percent annually. The mean is 0.7 percent annually.

ECONOMIC ANALYSIS

The purpose of this section is to determine the extent to which cost of the slaughter animal or, alternatively, cost per kilo of beef on a live animal basis can be reduced from technology adoption. The procedure is to apply the percentages calculated in Appendixes 8.1-8.4 to the 1983 costs and determine the costs, in constant 1983 yen, for the three scenario years. The beef breed costs, calculated on a total cash cost basis, are provided in Appendix 8.7. A summary of the totals, along with cost reduction on a percentage basis, are provided in Table 8.4. Results show that in 1990 the low estimate for cost reduction (as measured by cash costs) is 5 percent while the high scenario is for a 17 percent reduction. The 2010 estimates are 29 and 55 percent for the low and high scenarios respectively.

Table 8.4. Beef breed cost per slaughter animal in constant 1983 terms, Japan, 1990, 2000 and 2010

	Scenario		
Year	Medium	Low	High
	- - - - - - - - Yen - - - - - - - -		
Total cash cost			
1983	606,195	606,195	606,195
1990	536,159	575,194	500,587
2000	427,642	524,431	337,025
2010	347,320	428,409	272,562
	- - - - - Percent - - - - -		
Reduction in cash cost			
1990	12	05	17
2000	29	13	44
2010	43	29	55

Source: Appendix 8.7.

Computations of dairy breed net basic production costs (analogous to cash costs) provided in Appendix 8.8, and summarized in Table 8.5, point to a reduction of 7 percent in 1990 under the low scenario, and 13 percent under the high scenario. The reductions are 29 and 37 percent for 2010.

The economic analysis indicates that considerable potential exists for cost reduction, and that the majority of it can be obtained by 2000, i.e. a decade-and-a-half from 1985. In addition, scrutiny of the various technologies leads to a conclusion that the high estimates for 2010 can easily be attained by 2000 from a technical viewpoint. The serious drawback is human capital, i.e. the extent to which technology adoption and industry structural reorganization will take place.

Table 8.5. Dairy breed cost per slaughter animal in constant 1983 terms, Japan 1990, 2000 and 2010

Year	Scenario		
	Medium	Low	High
	- - - - - - - - Yen - - - - - - - -		
Total net basic production cost			
1983	397,163	397,163	397,163
1990	370,328	384,294	343,787
2000	307,179	340,871	273,677
2010	281,853	317,352	248,351
	- - - - - Percent - - - - -		
Reduction in net basic production cost			
1990	07	03	13
2000	23	14	31
2010	29	20	37

Source: Appendix 8.8.

COMPARATIVE STUDIES

The scenarios developed in this chapter can be placed in perspective by comparing them with a recent study carried out in the United States. Experts in the RCA Symposium titled "Future Agricultural Technology and Resource Conservation" (English, Maezold, Holding, and Heady, 1983) visualized meat production per breeding cow and per breeding sow in the United States increasing 25 percent by the year 2000, and 60 percent by the year 2030. Milk production per cow is expected to increase 40 percent by 2000, and 60 percent by 2030.

Additional projections from the RCA Symposium about livestock, poultry and fish production in the United States are that production per female sheep or goats will increase 35 percent by 2000 and 70 percent by 2030. Broilers, already the target of vast productivity improvements, are expected to experience a 30 percent improvement in production efficiency by the year 2000. Age-to-market weight efficiency of catfish is expected to increase 20 percent by the year 2000 and 200 percent by 2030.

The rather large productivity increases projected in the study are primarily the result of synergistic effects, i.e., that the total effect of two or more technologies in combination is greater than the sum of the individual ones. For example, the report indicates that a combination of embryo transfer and genetic engineering in the United States could increase milk yield up to 20,500 kilos per cow annually on selected dairies by the first part of the next century. We believe similar magnitudes are possible for Japan.

SUMMARY AND CONCLUSIONS

Scenarios for the year's 1990, 2000, and 2010, i.e. 5, 15, and 25 years from 1985 are developed in this chapter for beef production and productivity increases as well as cost reduction potential. The analyses are carried out by dividing beef production into four cattle classes, beef breed calves, dairy breed calves, beef breed feedlots and dairy breed feedlots. Beef production productivity, as measured by beef production per head of all cattle in the national inventory, is reckoned to reach 116 kg in 1990, 131 kg in 2000, and 140 kg in 2010 under the high estimate scenario. In contrast it was 109 kg in 1984. The low estimate scenario is for 110, 120 and 123 kg for the three respective periods. The growth rates for the two scenarios are 0.5 and 1.0 percent annually. In contrast, the rate was 3.7 percent from 1960 through 1984.

An economic analysis on the potential for cost reduction reveals that cost of animals being sold from beef breed feedlots will likely decline between 5 and 17 percent by 1990 (in constant 1983 monetary terms) and between 29 and 55 percent in 2010. Dairy breed cost for slaughter weight animals is calculated to decline between 3 and 13 percent in 1990 and between 20 and 39 percent in 2010.

The extent to which Japan can increase it competitive position vis-a-vis nations supplying her with beef depends greatly on the extent to which those countries are able to reduce production and marketing costs. Analysis of Japan's cattle industry reveals that more leeway potentially exists for cost reduction than in other countries because (1) high output prices provide incentive to introduce large amounts of inputs and new technology, and (2) the technologies evaluated in this chapter can more easily be adopted in the intensive production environment which characterizes Japan than in the more extensive type operations found in the United States, Australia and New Zealand. We now examine the impact from structural changes in Japan's cattle industry and then tie the supply side with the demand side to estimate beef imports under a number of scenarios.

9
Supply and Demand Side Scenarios

The potential impact on productivity and cost reduction resulting from technological and structural changes in Japan's beef industry was estimated in the last chapter. The problem now dealt with is to expand the scenarios by (1) developing additional ones about total supply of domestically produced beef, (2) amplifying the dicussion on dietary patterns presented in Chapter 2 into scenarios about total beef consumption and, (3) integrating supply and demand side scenarios to determine beef imports.

SUPPLY SIDE SCENARIOS: BEEF FROM DAIRY BREEDS

Domestically produced beef is a product of the national dairy and beef breed herds. As productivity increases, so will domestic beef supply (Figure 9.1). In addition, inventory adjustment will also bring about changes in domestically produced beef supply. The purpose of this section is to develop scenarios about domestically produced beef supply emanating from dairy cattle.

Much of the technology discussed in the last chapter will enhance milk production per cow. Japan's milk production is only about 82 percent that of the United States (4,633 kg versus 5,637 kg in 1982) and far less than in Israel, the world's leader in 1982 with 6,958 kg. Considering these factors plus technologies discussed in Chapter 7, it can be argued that Japan's average milk production per cow could easily reach 5,500 kg in 1990, 6,200 kg in 2000 and 7,000 kg in 2010. Per capita consumption of dairy products in Japan is still increasing, but at a decreasing rate. As a result, dairy production will increase faster than consumption which, in turn, means Japan will either

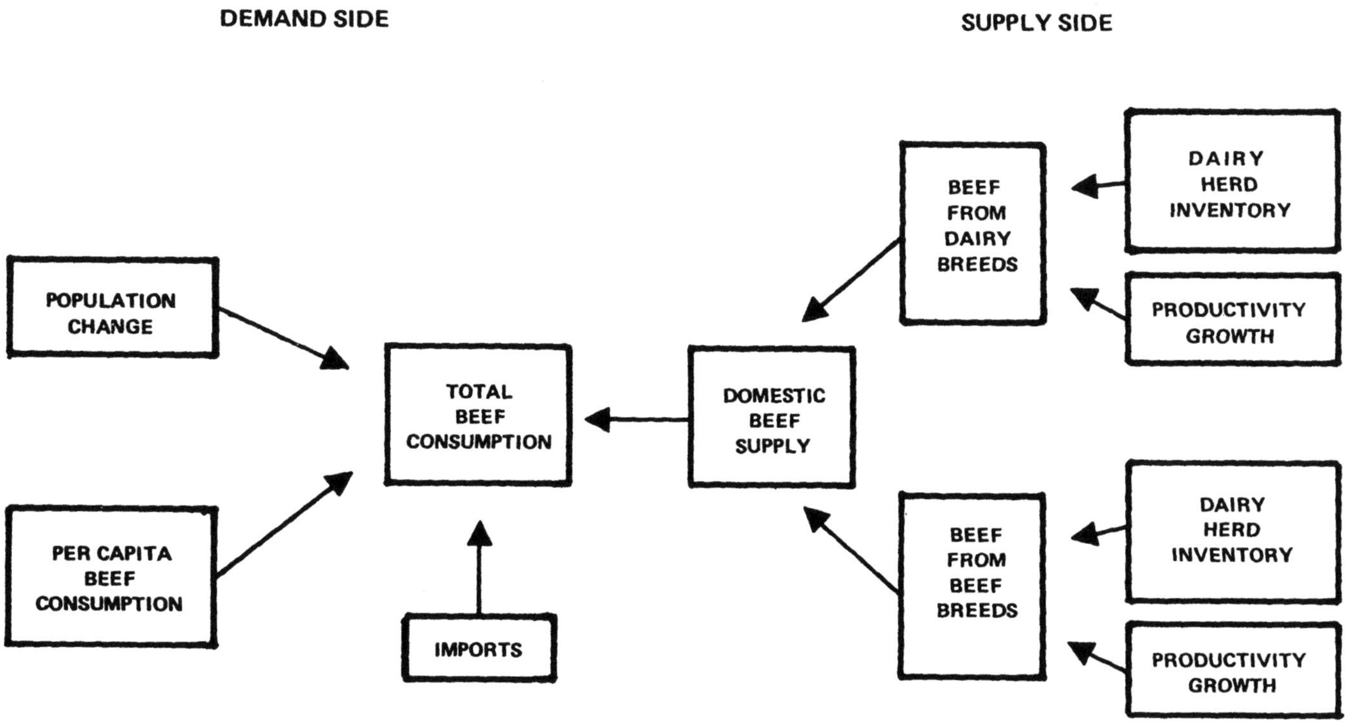

Figure 9.1 Demand and supply side interaction for beef production and consumption in Japan

have to begin exporting milk products (fairly unlikely) or reduce dairy cow inventory (most likely).

One scenario about the impact increased technology will have on production of dairy products is presented in Table 9.1. The analysis begins with recognition that there were 6,848,000 MT of domestically produced dairy products in 1983, or 57.55 kg on a per capita basis. The MAFF projects total dairy product consumption will be 73 kg per capita in 1990 which, if domestic production continued to be about 83 percent of consumption, would translate to 60.59 kg in 1990. We estimate a 5 percent increase from 1990 to 2000 and an additional 3 percent increase to 2010. That means consumption of domestically produced dairy products would be 63.02 kg in 2000 and 65.53 kg in 2010.

Examination of the aforementioned production per cow figures shows there would have to be a reduction of 7 percent in dairy cattle inventory by 1990, 9 percent reduction in 2000 and 13 percent in 2010 (compared with 1983) to maintain the 83 percent self-sufficiency level. These are used as our medium scenarios. Considerable leeway exists for argument about the assumptions in the simulation exercise developed in Table 9.1 but the results are fairly clear--dairy cow numbers will not likely increase in the future and, in all likelihood, will decline. That means dairy calf numbers would be reduced except to the extent to which a decline is offset by advances from technological change such as reduced death loss, multiple calving and other items quantitatively measured in Appendix 8.2.

The impact fewer dairy cows will have on total beef production is examined in Table 9.2 under three scenarios; one which results in the least imports, i.e. the most favorable to Japan, a medium one, and one which results in the most imports i.e. the least favorable to Japan. These three scenario categories will be used for the rest of this chapter. The first step in Table 9.2 is to determine beef and veal production from dairy cattle under the assumption that no technological change takes place. The percentages from Table 9.1 on dairy cattle inventory reduction are used to estimate beef and veal production from dairy cattle assuming no change in productivity. Then, the increased dairy calf productivity calculations developed in Appendix 8.2 are applied to yield the final production estimates.

The figures developed on the impact of technological change in dairy cattle feedlots (Appendix 8.4) are not included as the technologies do not change the total quantity of beef; rather, the impact is on the <u>rate</u> at which it is produced. Results for the three production periods under the least import scenario is 330 thousand tons in 1990 and 384 thousand tons in both 2000 and 2010. The most imports scenario is 319 thou-

Table 9.1. Milk cows in lactation required to meet domestically produced dairy product needs, Japan, 1990, 2000, 2010

Item	Year			
	1983	1990	2000	2010
Total population (millions, 0.49 percent growth rate)	119	123	129	136
Per capita consumption of domestically produced dairy products (kg)[a]	57.55	60.59	63.02	65.53
Total domestic dairy products production (1,000 MT)	6,848	7,453	8,207	8,912
Milk production per cow in lactation (kg)	4,700	5,500	6,200	7,000
Milk cows in lactation to meet domestically produced needs (millions head)[b]	1,457	1,355	1,324	1,273
Percent reduction from 1983	--	07	09	13

[a]Imports are about 17 percent of total dairy product consumption. MAFF projects total dairy product consumption at 73.00 kg in 1990 which means domestically produced product would be 60.59 kg by that year. Per capita consumption is assumed to increase 5 percent from 1990 to 2000, and 3 percent from 2000 to 2010.

[b]Cows 2 years and older.

Table 9.2. Beef and veal production from dairy cattle in Japan, 1990, 2000 and 2010

Item	Scenarios: Least imports	Medium imports	Most imports
Beef and veal production from dairy, 1983 (1,000 MT)	333	333	333
Projected reduction in dairy cows (Percent)[a]			
1990	09	07	05
2000	12	09	06
2010	17	13	09
Beef and veal production from dairy, no change in productivity (1,000 MT)			
1990	303	310	316
2000	293	303	313
2010	276	290	303
Increased dairy calf productivity (Percent)[b]			
1990	09	05	01
2000	31	22	13
2010	39	30	20
Beef and veal production from dairy, increased productivity, (1,000 MT)[c]			
1990	330	326	319
2000	384	370	354
2010	384	377	364

[a]Medium scenario from Table 9.1.

[b]From Appendix 8.2.

[c]Assumes feedlots make no additional contribution to total beef production.

sand tons of domestic production in 1990, 354 thousand tons in 2000 and 364 thousand tons in 2010.

Beef and veal production from beef breeds, along with imports, can be considered a residual supplier for analytical purposes since the dairy herd is relatively fixed. Also, beef production from beef breeds and beef imports are open to wide speculation due to a host of economic and political factors. The approach is to estimate consumption and then develop alternatives about the extent to which beef production from beef breeds will expand or contract.

DEMAND SIDE ANALYSIS

Consumer demand is the major force driving Japan's beef business. The several components which make up demand can be divided into two main categories; those which shift the demand curve, and those which cause a movement along the demand curve. Movements along the demand curve take place as a result of changes in supply or price. Domestic beef production has continued to increase over time which, if there were no shift in the demand curve, would have led to lower prices. Another source of supply is imports. Because exporters have lower production costs than Japan, imported beef should have the effect of reducing price at retail. But, because of quotas and a substantial import duty, it can be argued there has been little price effect. Liberalization of quotas, or substantial increases in them, combined with reduced import duties will lead to a movement down along the demand curve. That means lower prices and increased quantity taken.

There are a number of demand shifters. Three of them are changes in population, government policy, and in tastes and preferences. An example of the latter factor is introduction of new products such as hamburgers, which partially replace traditional ones like raw fish. Another demand shifter is change in income levels which result in so-called "superior" goods like meat or caviar replacing the so-called "inferior" goods such as root crops. Relative price of competing products also leads to shifts. For example, price of chicken as a percentage of beef price has declined due to rapid adoption of cost saving technology in chicken production, processing and distribution.

The demand curve for beef has tended to shift to the right in Japan as the expansion in tastes and preferences for beef, combined with income increases, have tended to offset the pull backward by the continually increasing price advantage of poultry. In other words, total quantity of beef has increased although price of beef, in real terms, has remained about constant.

Promotion efforts by domestic beef producers (mainly through LIPC) as well as export agencies from other countries such as the Meat Export Federation of the United States (MEF) or the Australian Livestock and Meat Commission, are aimed at increasing consumption per capita, i.e., shifting the demand curve for beef to the right. To date, foreign suppliers have been more interested in securing a greater quota for their product than simply expanding demand for beef.

Shifts in tastes and preferences are largely measured by income elasticities, with the implicit assumption being that as income increases "superior" goods replace "inferior" goods. But, since income elasticities are of necessity based on historical data, great care must be taken in using them for long-term projections since the rate in additional consumption of products tends to slow down as industries mature. We argue that consumption of meat in Japan is approaching maturity, albeit at a much lower absolute level than in the United States and a somewhat lower level than in Europe (total meat consumption in the United States is now 6.4 times that of Japan while the EC-10 consumes 3.9 times as much as Japan).

Income elasticity estimates vary considerably depending on the method used to estimate them and the period chosen. Yuize (1971) estimated income as 1.25 for the period 1951-1960. Monma (1984) estimated it at 1.00 using data from 1959-1981, while Kishimoto (1982) estimated it at 1.00-1.10 for the period 1970-1976. Kester (1980) projected it would be 1.30 for the period 1983-1985. Another researcher (Sanderson, 1978) projected it would be 1.89 for the period 1974-1985. The Organization for Economic Cooperation and Development (OECD) estimate for 1975-1985 (published in 1968) was 0.90. The Ministry of Agriculture in Japan (MAFF) uses 0.90 for beef consumption to 1990 while the U.S. Department of Agriculture's Economic Research Service (ERS) presents two scenarios in their projections, one using an income elasticity of 1.40 and the other 1.70. An extensive review of income and price elasticities is found in Coyle (1983).

We tend to be rather conservative in our projections of the impact which income increases will have on beef consumption. In addition to the arguments developed in Chapter 2, we give heavy weighting to realization that average age of Japanese is increasing. Ever greater proportions of the population in upper age cohorts means a slowdown in shifts to western diets because as people in Japan get older they reduce their beef consumption. For example, it is estimated that in 1982 females in the 30-39 age cohort consumed 4.02 kg per capita annually (retail weight for a selected group, i.e., not national averages) while those in the 50-60 age cohort consumed 2.83 kg and females 70-79 ate

1.09 kg (Table 9.3). Furthermore, slower gains in per capita income are likely for the rest of the century. As a consequence, we project the income elasticity to be 0.9 from 1983 to 1990 under the least import scenario, 1.0 in the medium scenario, and 1.1 in the most import scenario.

Considerable discussion could be generated about the impact of competing products (measured by cross price elasticities) and the extent to which quantity taken of beef will vary as a result of changes in price of beef itself (movements along the demand curve). Numerous studies (see Coyle, 1983) give conflicting results. A further source of confusion is that retail beef price increased faster than other meats from 1975-1980, but has increased much more slowly than pork since then and at about the same rate as chicken (Table 9.4).

There are no import quotas on pork or poultry. However, since October, 1971 pork imports have been controlled by a variable levy. The level of protection is determined by the standard import price (also called the minimum import price) which is midway between the floor and ceiling prices for carcass pork which, in turn, are established by the 1961 Law on Stabilization of Livestock Product Prices. The net result is imports are subject to the larger of a duty or levy equal to the difference between the C.I.F. import price and the standard import price. If the policy is encouragement of imports, the standard import price is replaced with proclaimed prices. The duty was lifted one time in 1972-1973, and led to dramatic upsurges in pork imports. In all likelihood the variable levy will be maintained to protect the domestic pork industry. Poultry imports are affected to some extent by tariffs which are about 20 percent ad valorem on chicken meat. The 1977 MTN agreement will reduce the tariff on chicken legs from 20 to 10 percent and on turkeys from 10 to 5 percent by 1987.

Given the uncertainty about the differential impact of cost reducing technological change in the industries of competing products and the uncertainities regarding policy, we assume the cross price elasticity for beef will be unitary. We also assume that price elasticity is unitary. Thus, income is the only variable used to measure shifts in the demand curve.

Real per capita private consumption expenditures increased at 4.5 percent annually from 1970-1975, but slowed down to 3.7 percent in the early 1980's. Our low or least import scenario is for it to average 2.5 percent from 1983-1990, and then 2.0 percent from 1990 to 2010 (Table 9.5). The medium estimate scenario is for 3.0 percent from 1985-2010 while the scenario which would result in the most beef imports is for real pri-

Table 9.3. Changes in beef consumption per capita in Japan, 1973-1982

Age cohort	Males			Females		
	1973	1982	Ratio	1973	1983	Ratio
	- - - - - Kg - - - - - -			- - - - - Kg - - - - -		
0-9	0.38	0.68	1.78	0.17	0.40	2.31
10-19	1.30	2.95	2.27	0.54	1.60	2.94
20-29	2.47	3.20	1.30	2.10	2.62	1.25
30-39	2.96	3.86	1.30	3.28	4.02	1.23
40-49	3.37	3.98	1.18	3.06	3.77	1.23
50-59	3.07	4.19	1.37	1.61	2.83	1.76
60-69	2.34	3.25	1.39	0.72	1.50	2.08
70-79	1.30	2.52	1.94	0.77	1.09	1.42

Source: Morishima, 1984.

Table 9.4. Index numbers of meat and consumer prices in Japan, 1975-1983

Year	Retail prices			Consumer Prices	
	Medium beef	Medium pork	Medium chicken	All commodities	Foodstuffs
	Index 1980 = 100				
1975	80.5	105.6	95.9	72.9	76.6
1976	88.3	109.4	105.3	79.5	83.6
1977	90.3	106.9	105.7	86.1	89.2
1978	89.9	104.2	99.0	89.4	92.3
1979	94.0	99.3	96.1	92.6	94.3
1980	100.0	100.0	100.0	100.0	100.0
1981	101.5	105.9	104.9	104.9	105.3
1982	103.5	110.0	105.5	107.7	107.2
1983	104.5	111.5	104.9	109.7	109.4

Source: Retail prices from MAFF. _Statistical Yearbook of Minstry of Agriculture, Forestry and Fisheries._ Tokyo, 1982 and 1985. Consumer prices from MAFF, _Abstract of Statistics on Agriculture Forestry and Fisheries,_ Tokyo, 1981 and 1984.

Table 9.5. Scenarios of per capita beef and veal consumption in Japan, 1990, 2000 and 2010

Item (calendar year basis)	Scenario: Least imports	Scenario: Medium imports	Scenario: Most imports
Historical per capita consumption (kg, carcass weight)[a]			
1983	5.8	5.8	5.8
Income elasticity of demand (%)			
1983-1990	0.9	1.0	1.1
1990-2000	0.7	0.8	0.9
2000-2010	0.2	0.5	0.8
Per capita real private consumption expenditure (%)[b]			
1970-1975	4.5	4.5	4.5
1982	4.2	4.2	4.2
1983	3.3	3.3	3.3
1984	4.2	4.2	4.2
1983-1990	2.5	3.0	4.0
1990-2000	2.0	3.0	4.0
2000-2010	2.0	3.0	4.0
Annual percentage change[c]			
1983-1990	2.25	3.00	4.40
1990-2000	1.40	2.40	3.60
2000-2010	0.40	1.50	3.20
Growth rate coefficient for the percentage change[d]			
1983-1990	1.18869	1.22987	1.35177
1990-2000	1.14916	1.26765	1.42429
2000-2010	1.04073	1.16054	1.37024
Per capita consumption (kg, carcass weight)[e]			
1990	6.89	7.13	7.84
2000	7.92	9.04	11.17
2010	8.24	10.49	15.31

[a]See Appendix 9.1. Note that the per capita shown in the Food Balance Sheet (Appendix 1.13) is on a retail weight basis. Also, the Japanese data is on a fiscal year basis. The data used are USDA since that is on a calendar year basis (Appendix 9.1).

[b]Complied by the Ministry of International Trade and Industry (MITI).

[c]Obtained by multiplying income elasticity of demand by per capita real private consumption expenditure.

[d]From growth rate tables.

[e]Growth rate times per capita consumption in new base year for each period.

vate consumption expenditures to increase at 4.0 percent annually.

The result of combining the various factors is a range in per capita beef and veal consumption in 1990 of 6.89 kg, 7.13 kg and 7.84 kg for the least import, medium, and most import scenarios, and 8.24, 10.49 and 15.31 kg in 2010 under the three respective scenarios (Table 9.5). The actual per capita consumption (these and all data presented are on a carcass weight and calendar year basis) in 1983 was 5.8 kg. The growth rate in per capita consumption was 5.7 percent annually from 1975-1983 and will likely be 4.4 percent from 1980 through 1985. Our Medium import scenario has a growth rate of 3.0 percent annually from 1983 through 1990 and 1.9 percent from 1990-2010. The most import scenario is 4.4 percent to 1990 and 3.4 percent to 2010. The next step is to develop scenarios of domestic beef production from beef breed imports.

INTEGRATED SUPPLY, CONSUMPTION AND IMPORT SCENARIOS

Two very different situations regarding production from beef breeds are now developed. The first one is an assumption that beef breed production does not change. In effect, even though production enhancing technologies are adopted, the political/economic climate is assumed to be such that production per head of inventory advances are offset by a decline in beef breed cattle inventory. Integration of the dairy breed production scenarios and the consumption scenarios reveal that under the least import scenario, self-sufficiency in beef production would fall from the 72 percent recorded in 1983 to 58 percent in 1990, 54 percent in 2000 and 51 percent in 2010 (Table 9.6). Under the most import scenario the self-sufficiency rate falls to 50, 36 and 26 percent in 1990, 2000 and 2010.

The situation developed above is interesting and useful, but much more realistic is recognition that the Government of Japan is committed to maintain beef self-sufficiency at approximately current levels. If a 72 percent level were maintained to 2010, and the previously developed dairy breed production and beef consumption estimates were to hold, then beef production from beef breeds in 1990 would have to be 1.73 times that of 1983, 2.14 in 2000 and 2.39 in 2010 under the least import scenario (Table 9.7). It would have to more than double (2.31) by 1990 under the most import scenario and would be 6.60 times as much by 2010 (compared with 1983).

SUMMARY AND DISCUSSION

A number of alternatives in addition to those presented in this chapter could be evaluated such as milk production per dairy cow increasing at a rate other than the one assumed, or consumption of domestically produced dairy products increasing beyond the 83 percent level. But, the two situations provide sufficient information to develop some reasonable estimates about what will happen regarding beef production and imports. First, the Government of Japan has decreed it will place great emphasis on maintaining beef production at approximately current levels of self-sufficiency (about 72 percent). Given past successes in meeting targets for other commodities, the situation depicted in Table 9.7 would seem more realistic than the one in Table 9.6 where beef production from beef breeds remains stagnant and imports increase very rapidly.

We believe that the medium scenario in Table 9.7 is most likely. That means imports would increase from the 196 thousand tons recorded in calendar year 1983 to 246 thousand tons in 1990, 324 thousand tons in 2000 and 382 thousand tons in 2010. That represents an increase in imports of 26 percent to 1990, 65 percent to 2000 and 95 percent by 2010. If the medium scenario does prevail then beef production from beef breeds would have to increase 88 percent by 1990 compared with 1983. It would have to increase 186 percent over the 17 year period from 1983 to 2000 (i.e. be 2.86 times that of the 1983 level) and increase 273 percent over the 27 year period from 1983-2010.

The most basic factor to determine if beef breed production can increase substantially is economics. The budgets in Chapter 6 indicate that while substantial cost reduction will be achieved in the longer term, it must be recognized that other countries will also adopt cost reducing technologies. But, there is one technology which deserves more consideration--embyro transfer mainly for multiple calving. As indicated in Chapter 7 we believe the potential--from a technical viewpoint--will exist by 1990 to increase the number of calves from the national inventory (in both dairy and beef breeds) 25-35 percent by multiple calving through embryo transfer. That means total beef production could increase about the same amount. A 25 percent increase in beef production would essentially eliminate the need for additional imports. Is this possibility reasonable? We believe so despite the very conservative estimates in Appendixes 8.1 and 8.3, and cite the case of the EEC as an example. Fifteen years ago, and even a decade ago, it appeared that the EEC would continue to be a major beef importer. They are

Table 9.6. Beef and veal imports in Japan assuming no change in beef breed production, 1990, 2000 and 2010

Item (calendar year basis	Scenario		
	Least imports	Medium imports	Most imports
	- - - - - - - 1,000 MT - - - - -		
Beef breed production			
1983-2010	162	162	162
Dairy breed production[a]			
1983	333	333	333
1990	330	326	319
2000	384	370	354
2010	384	377	364
Total production			
1983	495	495	495
1990	492	488	481
2000	546	532	516
2010	546	539	526
Consumption[b]			
1983	691	691	691
1990	847	877	964
2000	1,014	1,157	1,430
2010	1,071	1,364	1,990
Imports			
1983	196	196	196
1990	355	389	483
2000	495	625	914
2010	525	825	1,464
	- - - - - - Percent - - - - - -		
Self-sufficiency			
1983	72	72	72
1990	58	56	50
2000	54	46	36
2010	51	40	26

[a]From Table 9.2.

[b]Per capita consumption in Table 9.4 multiplied by population of 123 million in 1990, 128 million in 2000 and 130 million in 2010. 1983 from Appendix 9.1.

Table 9.7. Beef breed production required to maintain a 72 percent beef and veal self-sufficiency level, Japan, 1990, 2000 and 2010

Item (calendar year basis)	Scenario: Least imports	Medium imports	Most imports
	Percent		
Self-sufficiency 1983-2010	72	72	72
Consumption[a]	1,000 MT		
1983	691	691	691
1990	847	877	964
2000	1,014	1,157	1,430
2010	1,071	1,364	1,990
Total production (72 percent self-sufficiency)[b]			
1983	495	495	495
1990	610	631	694
2000	730	833	1,030
2010	771	982	1,433
Imports			
1983	196	196	196
1990	237	246	270
2000	284	324	400
2010	300	382	557
Dairy breed production[a]			
1983	333	333	333
1990	330	326	319
2000	384	370	354
2010	384	377	364
Beef breed production[c]			
1983	162	162	162
1990	280	305	375
2000	346	463	676
2010	387	605	1,069
Ratio, beef breed production over 1983	Ratio		
1990	1.73	1.88	2.31
2000	2.14	2.86	4.17
2010	2.39	3.73	6.00

[a]See Table 9.6.
[b]72 percent of consumption.
[c]Difference between total production and dairy breed production.

now a close second to Australia, the leading net exporter. That change in position came about as a result of technology and policy.

Another factor to consider about embyro transfer is economic benefit to producers. Embyro transfer has mainly been justified in the United States on the basis of genetic improvement enhancement. But, the situation in Japan is much different since production cost of calves, and price of them, is much higher than in the United States. If a beef breed calf production cost of $1,200 is assumed as a base, additional transfer related cost estimated at $400, and calf crop doubled to 1.6 calves per cow from 0.8 today, then cost per calf would be $1,000. That is a 17 percent reduction in cost. Clearly, the benefit to dairy producers would be lower, but it is also expected that by 1990 transfer costs will be reduced substantially.

Overall, we believe there is a remarkable potential to expand inventory as well as improve productivity per animal. Embryo transfer for multiple births will permit considerable total inventory increase without changing the number of mature animals. Decreases in dairy breed inventory will release land and other resources for beef breed use. The rice land diversion and marginal land use programs will provide not only land but also additional resources for beef cattle production. Finally, the rice diversion program combined with research on more efficient use of rice straw and development of large scale dry lot cow/calf operations will lead to expanded inventory in agricultural regions which are now relatively minor beef producers.

Our scenarios about the impact of technological change (Chapters 7 and 8) are quite conservative. This is because effort has been to provide a framework for analysis and we have strived to provide considerable detail about individual technologies so that alternative scenarios can easily be developed. It must be cautioned again that these exercises, just like the previous ones, are not forecasts or even projections. Rather, the exercises are "what-if" scenarios for analytical purposes. The one last piece of information required is discussion and estimates on government policy. That is the topic of our last chapter.

10 Ramifications and Policy Implications of New Technologies

A myriad of technologies now used in Japanese beef production, and ones that will be available in the next 25 years, have been evaluated and various scenarios developed about the impact on industry structure. These supply side phenomena have been amalgamated with demand side data, and various other scenarios developed about the impact on the country's beef consumption, production and imports. The purpose of this final chapter is to provide an overview of policy issues that arise from the scenarios, government policy goals regarding beef production, and an evaluation of current and projected policies.

GOVERNMENT BEEF INDUSTRY POLICIES: AN OVERVIEW

The Japanese government's basic aims regarding beef have been (1) to assure that producers have an "adequate" total income level and (2) to enhance the country's self-sufficiency in beef production. Overall, it appears that while the first goal has been relatively successful, despite a major exodus of producers from the cattle business and fairly rapid increases in farm size, the second policy goal of maintaining or improving self-sufficiency in beef production has failed since the ratio declined from 96 percent in 1960-1965 to about 72 percent in recent years.

It is true that serious impediments to expanded beef production exist such as limited pasture, and little grain and forage production land. Nevertheless, analysis of beef production technologies indicates that a very low adoption rate of them by Japanese farmers is the major constraint to expanded beef production rather than lack of land. In a word, while impressive progress has been made in poultry, swine and crop production, beef production from beef breeds is relatively inefficient.

Beef production per head of inventory is high compared with other developed countries, but is due to nearly three quarters of beef being produced from single purpose dairy cattle and not by contribution from beef breeds. Furthermore, large gains in that productivity measure have mainly been due to development of a commercial feedlot industry and, in recent years, a shift to longer feeding periods. Thus, ironically, what appears to be an improvement in productivity actually masks a certain amount of less efficient resource use.

Beef breed production inefficiency is partially a result of very protective government policies, but ones promulgated for very important reasons. One justification is great concern about maintaining a secure and stable beef supply since this commodity has increasingly become a major protein source. An example of apprehension about domestic production versus imports is the emphasis on self-sufficiency ratios in Japan. This measure is virtually unheard of in the United States except for strategic defense related materials such as certain metals and petroleum.

However difficult, western analysts must realize that Japanese view food as a strategic area because of the country's historical and geographic position. Food shortages during and after World War II are still vividly remembered and, while American's might dismiss the soybean embargo of 1973 as an insignificant tactical error by the U.S. administration, that apparently minor incident is well remembered by Japanese policy makers and continually serves to reinforce awareness of the need for setting minimal self-sufficiency levels. Our analysis (medium scenario) indicates that total beef consumption will grow from about 671 thousand MT in 1983 to 877 thousand MT in 1990 and 1.2 million MT by 2000. Thus, if a minimal level of 72 percent self-sufficiency is to be maintained (a generally accepted figure in Japan), domestic beef production must increase 27 percent between 1983 and 1990 and 68 percent between 1983 and 2000.

A second reason the Japanese government has continued to pursue quite protective policies regarding beef is that production of that commodity is a relatively high land using type of production. Japan is limited in the marginal land typically used in beef production. There is marginal land available which basically has no alternative farming use, but development of it even for beef production is a high cost proposition. However, there is a national desire to use this land as a means to provide employment for as many families (including older people) as possible to sustain outlying disadvantaged regional economies. In effect, land use has been closely tied to social policies.

A third factor is the close linkage of beef production with the dairy sector which, like the dairy industries in the United States and the EEC, has been highly protected. In Japan nearly 70 percent of domestic beef is derived from dairy cattle, and thus the close ties are inevitable.

Protectionistic policies have their strong points, but they are also partially responsible for the widespread inefficiencies and resultant high cost in beef production. The basic problems now facing policymakers are (1) how to reduce production cost (2) the means to improve industry structure and (3) the mechanism to attract progressive and active beef cattle operators. We now turn to a discussion of the policy instruments which have been used.

POLICY INSTRUMENTS

The government of Japan has pursued a number of protective price and subsidy policies. These fall under the categories of import quotas, domestic beef price stabilization policies, calf support price policies, production and structural improvement policies, and service related ones.

Import Quota Policies

Agricultural trade liberalization started in the mid-1960's concurrent with the takeoff in Japan's economy. At that time there were 73 items restricted. In the early 1970's quotas were either removed or relaxed on most of them so that as of early 1985 there were only 22 non-liberalized agricultural items, among which beef and oranges are dominant.

Import controls on beef were first imposed in the late 1950's in an effort to lessen the drain on scarce foreign exchange. But, the most important steps were taken when the Livestock Industry Promotion Corporation (LIPC) was created in 1961 to carry out the Law for Price Stabilization of Livestock Products. Further momentum was achieved between 1965 and 1971 when LIPC and certain designated private traders were allocated the beef import quota. Gradually, LIPC obtained a larger share of the total quota and now has 90 percent of it. However, that share is now in the process of being reduced.

The import quota system has gradually changed to include more flexible schemes during the past decade. For example, in 1970 the "one-touch" system was adopted which allows authorized importers to sell LIPC quota beef directly to designated distributors. The purpose is to reduce handling cost of chilled beef. On the other hand, a more restrictive "tender" system was introduced in 1971 to insure that quality and quantity of frozen imported beef meet predetermined decisions about the desirable kind of imported beef. LIPC periodically issues tenders, obtains the commodity and then sells it to end users by auction at wholesale markets.

Japan imposes a 25 percent ad valorem tariff on all imported beef. In addition, various surcharges are also levied to reduce the disparity between imported and domestic beef prices. Revenue obtained from the above instruments is used by LIPC to support and promote the livestock industry.

Domestic Beef Price Stabilization Policy Through LIPC

LIPC has stabilized and controlled beef prices through stock adjustment and the semi-annual adjustment of the import quota ever since the system was initiated in May, 1975. The mechanism is to maintain wholesale prices of popular grade Wagyu and dairy steer beef at predetermined levels. The mid-point (medium) price is calculated based on the previous 7 years wholesale prices and estimates of current fattening costs. Floor and ceiling prices are then calculated on the basis of historical price variation. Dairy steer beef prices are also set based on past prices. Japan's beef prices, even in nominal terms, are very stable when compared with other OECD country's prices.

Feeder Calf and Feed Price Support Policy

A feeder calf price support policy was introduced during the 1960's to help alleviate a general decline in Wagyu cattle numbers. The approach followed is to give cow/calf producers a subsidy when feeder calf prices fall below predetermined levels. A stabilization fund (subsidy), managed by prefectural and national governments as well as producers, obtains about 50 percent of its financing from either the central government (MAFF) or LIPC.

A government managed feed price stabilization fund is also used to stabilize feed price. It basically operates through a mechanism of stockpiling feed grains.

Production and Structural Improvement Policies

The government of Japan has attempted to reduce beef production cost and modernize the beef industry by a host of mechanisms. One is a subsidy system which has not only been fairly high cost, but also quite ineffective mainly because of it being so specific. For example, much of the subsidy has been low interest loans with very tight specifications about type of facility, standards and other technical details which have often been inappropriate for the individual farmer conditions.

Another subsidy target is for pasture enlargement and improvement. Public pasture lands have been developed, but all of these efforts have been very high cost. There have also been subsidies paid to promote double-cropping of hay and other fodder crops on rice paddy land, and as a complement to the paddy land diversion programs. However, these programs have only been marginally successful and have been high cost.

The system of subsidies and loans has been less than satisfactory because it has often resulted in an excessive financial burden on producers despite the relatively low interest rates. Many "modernized" producers have suffered severe financial difficulty as a result of being forced to adopt equipment and facilities that were too expensive for their cash flow situation. In effect, attempts have been made to _push_ development by financial means rather than _pulling_ it through a coordinated effort based on careful analysis of debt structuring and management improvement.

Analysis shows that many of the structural policies, including land consolidation and improvement, have been intended to bring about rapid expansion in farm size. Tied to this has been farm legislation modified in 1975 and again in 1980 which relaxed WW II Occupation Force promulgated limits on size of holdings. But, even though average farm size has continued to grow, the vast majority of farms have not expanded to the extent envisioned by policy makers who designed complementary subsidy and loan policies. The result is most Japanese farm households have combined farming with off-farm employment. In essence, due to rather limited opportunities for farm size expansion, part-time farming is considered a necessity rather than a choice for most Japanese farm families.

Other Beef Related Policies

Government provides only limited extension and farm management services, and they are relatively ineffective. In fact, most extension type services are obtained through the private sector, primarily via cooperatives. Extension service activities are carried out by prefectural governments on a model of the U.S. agricultural services. But, unlike the U.S. land grant system which includes teaching, research and extension, each of Japan's activities is carried out by a separate body; teaching at universities, research at experiment stations and extension by prefectural governments. However, there is close cooperation between research and extension as both are under the auspices of the same departments in the prefectural government. There is definitely a missing link between research and farmers, and it does not appear that policies have been or are being fostered to overcome this knowledge gap.

One analytical and applied approach which can be expected to receive much attention as a methodology for beef breed cow/calf enterprise development is the farming systems research and extension (FSR/E) methodology. Principles of this technique, which was conceived in third world development circles for use with small, limited resource farmers, is now being widely used in the United States and is receiving increasing attention in other OECD countries. It is an approach which is different from traditional farm management research and extension in that it is (1) oriented towards the needs of small farmers, (2) specifically sets priorities for research which reflect the holistic perspective of the whole farm/rural household and natural and human environments, and (3) is aimed at defining general farming systems in a region under study as a means to identify and describe a population's activities. It is also more of a social institution oriented approach than farm management because greater recognition is being given to changes in non-technical factors such as markets, pricing policy and infrastructure. Another characteristic is an interdisciplinary orientation.

There have been remarkable changes in beef marketing but great inefficiencies still exist, mainly in the central wholesale market system. Government has not taken a vigorous approach to completely shift over to direct beef marketing using the system of regional packing plants which has been established. In fact, many plants, although very modern, are quite high cost largely due to low use of capacity. Other policy problems also exist which cause confusion among producers, such as lack of decision about a new grading system.

Overall, Japan's approach to beef industry policy has essentially been a supply side phenomena, and only recently has there been more demand side influence. This change is taking place mainly in response to greater health consciousness among consumers. The result is that even though beef consumption has become much more differentiated as per capita consumption has increased, the trade (meaning producers and wholesalers) have insisted on targeting production toward heavily marbled beef.

IMPACT OF U.S. PRESSURE FOR BEEF TRADE LIBERALIZATION

Japan's trade surplus with the United States, a chronic trade deficit by the United States, and emerging U.S. cattle industry related problems such as a decline in per capita consumption of beef, attendant low prices, and financial hardship by U.S. cattle producers have led the United States to press for Japan to liberalize beef imports. Japan has reacted cautiously, providing as concessions a special high quality beef quota for the United States during the Tokyo Round of GATT negotiations, an expansion of the quota in 1984, and various non-tariff barrier adjustments such as greater access by U.S. packers directly with end users.

The pressure to liberalize beef imports along with the possibility of lower tariff rates on imported beef, and another round of talks about beef trade liberalization which will begin in 1986, has provided considerable impetus by government officials to determine ways in which Japanese beef production cost can be dramatically reduced so that domestic producers will be able to compete. It is possible for the government to provide greater feeder calf subsidies as a means of competition, but there is great pressure to avoid this strategy given national priorities on budget austerity. In addition, there is great reluctance for Japan to fall into the EEC type subsidy trap.

Oddly enough, it appears that with the differentiation in beef consumption taking place, countries other than the United States such as Australia and New Zealand might be the main beneficiaries. The United States now accounts for nearly 30 percent of beef imports compared with only 9 percent in 1977 (Appendixes 10.1 and 10.2). In addition, there is great likelihood that if beef were completely liberalized EEC traders would attempt to penetrate this market given the heavy oversupply of beef in that common market. Regardless of the outcome on trade negotiations one thing is clear--the United States posture has led to expanded Japanese policy emphasis to reduce production cost and to expand output.

FUTURE BEEF POLICY

Institutions are the rules of a society or of organizations that facilitate coordination among people. They play a crucial role in determining resource use. The optimum, from an economic efficiency viewpoint, it to have institutions which provide order and stability, but are sufficiently flexible to permit change in response to endogenous and exogenous stimuli. As economic relations (input/output price, i.e. factor/ product price ratios) move into significant disequilibra, then institutional changes are required. There is, of course, always a struggle between vested interest groups, and between tradition and progress (however that may be defined). Innovations are thus influenced by the cost of achieving social consensus. Ruttan and Hayami's insistance that institutional change be treated as endogenous to the economic system represents a clear departure from modern analytical economics. But, in our opinion, the endogenizing of institutional change is crucial to understand and project technology adoption in Japan's beef industry.

Japan's history of agricultural development has been characterized by adoption of biotechnology which has permitted rapid growth in output despite severe land constraints. In contrast, United States growth in agriculture has primarily been characterized by mechanization, a means of increasing the area operated per worker. In Japan, it has not been profitable to substitute land for labor, and mechanical power has only partially been substituted for labor. In both countries factors of production have been substituted for each along what Ruttan and Hayami call a metaproduction function, i.e. an envelope of available production functions at different points in time. The substitution process has primarily been in response to long-run trends in factor prices. Technology in both countries has been directed to saving the relatively scarce factors of production--ones which have been radically different in each country. The important point is that the production fuctions have changed to conform with long-term trends in resource endowments and factor prices. The success in each country seems to lie in the capacity of producers, research institutions and agroindustries to exploit new opportunities in response to information transmitted through relative price changes.

There are a number of factors which will influence future beef policy in Japan apart from output prices. On the supply side our analysis indicates there is an extensive amount of technology available which could lead to substantial cost reduction. Of importance is that the technologies fit in the Japanese environment because they are mainly intensive type ones. But, regardless of the potential, it appears that adoption of them will be relatively low due to management constraints. Farm size is small thus compounding the adoption and service problem. Dairy cattle have provided nearly 70 percent of beef, but cattle inventory in that sector will likely decline thus placing more burden on production from beef breeds--and that is the sector which is the least efficient and the one which is the hardest to improve. Regardless, any moves to liberalize beef imports will lead to more competition and a "weeding out" of less productive producers.

On the demand side, analysis demonstrates that total beef consumption will increase but the greatest change in demand will be toward more popular (or medium) quality beef which is tender, but not highly marbled. The dietary pattern has essentially matured.

Review of the climate in which policy is set indicates there will be pressure to reduce rather than expand producer price and subsidy supports. The result will be redirection to more use of loans to stimulate production. It is also expected that funds will be targeted to the most viable and progressive producers. Emphasis will be on promoting farm size to reduce production cost through size economies. In effect, there will be a shift in emphasis from price to structural policy all linked with technology adoption.

There will be more attention given to resource reallocation to expand beef production. Most feasible is expanded use of land from rice diversion programs. Japan's rice price support program has resulted in domestic producer support prices being between three and four times the price of rice in international markets. The result is obstruction of resource shifts from production of rice to other commodities such as livestock. The government has also been forced to adopt a very expensive rice acreage control program.

Technological change, according to Ruttan and Hayami, is induced not only at the producer level, but at the national level as well. To them, there is a dialectic interaction between producers (through their organizations) and public sector research and extension organizatons. The more decentralized the system, the more responsive it is to demand by the producer clientele. In their public sector induced innovation model, research group response represents a critical link in the inducement mechanism. Price relationships play the lead role in deciding much of the pressure placed on research groups regarding relative priorities which society places on research goals. The public sector induced innovation model can be extended to explain why there will be some shift in priorities to greater beef cattle research.

Part of the new research and policy prioritization will include greater use of rice straw for roughage and expanded emphasis on development of new forage sources. For the longer run, if tenure and cost problems can be overcome, there will be greater policy emphasis on use of marginal resources such as mountain and foothill land. Finally, research and extension efforts will be redirected from crop production, where considerable successes have already been achieved, to beef and forage production. Tied with this will be greater weight placed on linking crop and beef production at the community and regional levels.

New techniques are not substitutes for labor or land per se. Rather, they are catalysts to facilitate substitution of relatively scarce factors of production for less scarce factors. Land in Japan is a relatively scarce production factor so this constraint can mainly be offset by the various advances in technology which increase yield per unit of land. This process of technological change (following Ruttan and Hayami's theory of induced innovation) should be considered endogeneous to the system rather than as exogenous. In their view technical change represents a dynamic response to changes in resource endowments and to growth in demand. Thus, a fully developed equilibrium theory of induced innovation should incorporate the mechanisms by which both product demand and factor endowments interact to influence the rate and direction of technological change.

Our view is that reliance will be placed on private sector initiative. Thus, whereas government has played a major role in the cattle industry, it is likely that emphasis will shift toward more reliance on the market place to allocate scarce resources. Combined with that will be less weight given to protecting small producers and more on viewing the cattle industry as a business rather than a social problem. That does not mean there will be sudden shifts, but rather that a trend will develop.

A second likely and even more important area for policy priorities will be on technology development and information dissemination. The problem will be how to foster and secure progressive beef operators who will be the most active adoptors of new technology. The approach will likely be expanded attention on selected persons and groups who will then be provided the financial and other aids necessary. Regional differences will be emphasized to a greater extent, i.e. beef production will be recognized as a heterogeneous business rather than as a homogeneous enterprise. Misunderstanding about this attribute of beef production has been a serious constraint to realization of cost reduction goals.

New legislation to allow greater use of underutilized upland resources, especially forest and nationally owned land for beef production, is a third likely area. Tied with this will be policies to make greater use of diverted paddy land. It is expected that large scale drylot cow/calf units not currently found in Japan (100 to 200 head in the longer run) will be promoted in areas with large amounts of diverted paddy land. These operations will most likely be cooperatively owned by farmers of the diverted land. They would then hire one or two persons to take care of the feedlot. Analysis indicates that with embryo transfer for multiple calving, economies from low labor requirements, mechanization and other technological and management advances, cost per beef breed calf sold could be reduced by at least 40-60 percent. Part of this analysis takes into account cooperators providing forage on the basis of cash costs only (they already own the land and most associated machinery) and availability of relatively inexpensive rice straw (which is mostly unused today).

A fourth area due for policy change is mobilization of land use rights based on rental or lease arrangements to stimulate land consolidation as a means to reduce costs. Tied to this will be a fifth policy target of human capital development. The problem is small size in beef breed cow/calf operations leads to low return on investment as well as only modest total income. As a result, progressive, innovative people are not attracted to this business. As a conseqence, a vicious circle prevails which prevents industry improvement. Recognition of human capital as the major constraint will lead to a major push to develop technology transfer mechanisms. Included will be expanded training of extension specialists and widespread research and information dissemination on record keeping and analysis.

SUMMARY AND FINAL REMARKS

Japan's beef business is widely misunderstood, not only by foreigners but even by Japanese well connected with the industry. One reason is the rapid structural changes which have been taking place and will continue to take place. The beef industry is much more efficient than a decade ago, but is still an island of traditionalism compared with pork and poultry production. We believe that much of the reason for the beef industry changing more slowly than the pork or poultry sectors has been slowness by the industry to recognize the rapid differentiation of beef consumption taking place. There has also been an attitude that the beef breed part of the industry should be considered as a social rather than economic subsector.

We initiated this book with a hypothesis (set forth in the preface) that beef production by Japan's cattle industry could be competitive with imported beef. Our conclusion is that while the technologies will be readily available by 1990 to support this hypothesis for medium to upper quality beef, adoption will be slow and domestic cost will not be reduced to the extent that Japan could compete directly with beef imports from the United States, and certainly not from Australia.

Another hypothesis stated in the preface is that Japan's beef industry has more potential for development than is generally recognized. We feel our research supports this hypothesis. Our analysis shows that while human capital and institutions have been a main bottleneck to cost reduction, these constraints can and will be overcome as industry structural changes continue to take place. This finding provides a new dimension to understand livestock industry development within Japan's general economy.

English Bibliography

Allen, C.E., D.C. Beitz, D.A. Cramer and R.G. Kauffman. Biology of Fat in Meat Animals. NC Pub. 234. Res. Div. College of Agriculture and Life Sciences, University of Wisconsin, December, 1976.

Anderson, Kym. "The Peculiar Rationality of Beef Import Quotas in Japan." Amer. J. Agr. Econ. 65(1983): 108-112.

Baker, Robert D. and Brain F. Shea. "Commercial Splitting of Bovine Embryos." Theriogenology 23 (January, 1985):3-12.

Bale, Malcolm D. and Bruce L. Greenshields. "Japanese Agricultural Distortions and Their Welfare Value." Amer. J. Agr. Econ. 60(1978):59-64.

Barker, J.S.F., Keith Hammond and A.E. McClintock. Future Developments in the Genetic Improvement of Animals. New York: Academic Press, 1982.

Berardi, Gigi M. and Charles C. Geisler. The Social Consequences and Challenges of New Agricultural Technologies. Boulder, Colorado: Westview Press, 1984.

Blair, Jess. "He'd Rather Feed Holsteins!" Beef, February, 1983, p. 24.

Bonnen, James T. "Historical Sources of U.S. Agricultural Productivity: Implications for R&D Policy and Social Science Research." Amer. J. Agr. Econ. 65(1983):959-966.

Bull, Alan T., Geoffrey Holt and Malcolm D. Lilly. Biotechnology: International Trends and Perspectives. Paris: OECD, 1982.

Burroughs, Wise, et al. Summary of Three Experiments: Feedlot Performance, Marketing Considerations and Carcass Characteristics of Holstein vs. Beef-Breed Feeder Steers. A.S. Leaflet R 70. Iowa State University, June 1965.

Butler, L.J. and A. Allen Schmid. "Genetic Engineering and the Future of the Farm and Food System in the U.S." in The Farm and Food System in Transition: Emerging Policy Issues. Michigan State University, 1983.

Cameron, John. "The EEC Beef and Veal Regime and its Implications for the Australian Beef Industry." Unpublished paper presented at the 28th Annual Conference of the Australian Agricultural Economics Society, University of Sydney, 7-9 February, 1984.

Castle, Emery N., and Kenzo Hemmi, eds. U.S.-Japan Agricultural Trade Relations. Baltimore: The Johns Hopkins University Press, 1982.

Coyle, William T. Japan's Feed-Livestock Economy: Prospects for the 1980's. Washington, D.C.: USDA FAER 177, February, 1983.

English, Burton C., James A. Maetzold, Brian R. Holding, and Earl O. Heady. RCA Symposium: Future Agricultural Technology and Resource Conservation, Executive Summary. Ames, Iowa: Center for Agricultural and Rural Development, 1983.

Fotenot, J.P. "Present Status and Future Trends in Production of Red Meat, Dairy, Poultry, and Fish with Emphasis on Feeding and Nutrition." Paper at the RCA Symposium: Future Agricultural Technology and Resource Conservation, Washington, D.C., 5-9 December, 1982.

Francks, Penelope. Technology and Agricultural Development in Pre-War Japan. New Haven: Yale University Press, 1984.

Gillian, Henry C. The U.S. Beef Cow-Calf Industry. Washington, D.C.: USDA AER No. 515, September, 1984.

Gregory, Keith E. "Breeding and Production of Beef to Optimize Production Efficiency, Retail Product Percentage and Palatability Characteristics." J. An. Sci. 55(1982):716-726.

Hayami, Yujiro. "Trade Benefits to All: A Design of the Beef Import Liberalization in Japan." Amer. J. Agr. Econ. 61(1979):342-347.

Hayami, Yujiro and Vernon E. Rutton. Agricultural Development: An International Perspective. Baltimore: The Johns Hopkins University Press, 1985.

Heady, Earl O. and Shashanka Bhide. Livestock Response Functions. Ames: Iowa State University Press, 1984.

Hedrick, H.B. Bovine Growth and Composition. Res. Bull. 928. Agr. Exp. Sta., University of Missouri, January, 1968.

International Monetary Fund (IMF). International Financial Statistics. Washington, monthly.

Iritani, A. "Present Status of Bovine Embryo Transfer in Japan." Embryo Transfer Newsletter (3)2,1984: 21-23.

Japan (Government of). "Basic Policies to Promote the Modernization of Dairy Farming and Beef Cattle Farming." Translated by the Agricultural Policy Research Committee, Inc., Tokyo, March, 1984.

____________________. "Law Concerning the Price Stabilization of Livestock Products and Other Matters." Translated by the Agricultural Policy Research Committee, Inc., Tokyo, March, 1984.

____________________. "Law Concerning the Promotion of Dairy Farming and Beef Cattle Farming." Translated by the Agricultural Policy Research Committee, Inc., Tokyo, March, 1984.

____________________. "Law Concerning Temporary Measures with Regard to Deficiency Payments of Milk for Processing to Producers and Other Matters." Translated by the Agricultural Policy Research Committee, Inc., Tokyo, March, 1984.

Gregory, K.E. and L.V. Cundiff. "Crossbreeding in Beef Cattle: Evaluation of Systems." J. An. Sci. 51(1980):1224-1242.

Jeremiah, L.E. A Review of Factors Affecting Meat Quality. Technical Bulletin 1. Research Station, Lacombe, Alberta, 1978.

Kada, Ryohei. Part-time Family Farming. Tokyo: Center for Academic Publications, 1980.

Kalter, Robert J., et al. Biotechnology and the Dairy Industry: Production Costs and Commercial Potential of the Bovine Growth Horome. AE Research 84-22, Department of Agricultural Economics, Cornell University, December, 1984.

Kester, Annie Y. "Demand for Fish in Japan and the Westernization of the Japanese Diet." Washington, D.C.: USDA ESCS. Unpublished manuscript, May, 1980.

Kester, Warren. "New Way to Feed Holsteins: From Birth to Market on High-Energy, Low-Roughage Rations." Beef, March, 1983, pp. 61-62, 64, 66, 99.

King, W.A. "Sexing Embryos by Cytological Methods." Theriogenology 21(January, 1984):7-17.

Livestock Industry Promotion Corporation (LIPC). Japan's Beef Figures. Tokyo, monthly.

__________________. Weekly Livestock Market Report. Tokyo, weekly.

Longmire, Jim and Walter H. Gardiner. Long-Term Developments in Trade in Feeds and Livestock Products. Washington, D.C.: USDA FAER 199, January, 1984.

Longworth, John W. and Ryohei Kada. "Japanese Food and Feedstuffs Imports: Background and Prospects." Dept. of Agr., Agr. Econ. Discussion Paper 1/82, University of Queensland, Australia, revised, June, 1982.

Longworth, John W. Beef in Japan. New York: University of Queensland Press, 1983.

Lu, Yao-Chi, Philip Cline and Leroy Quance. Prospects for Productivity Growth in U.S. Agriculture. Washington, D.C.: USDA ESCS Agr. Econ. Rpt. 435, September, 1979.

Ministry of Agriculture, Forestry and Fisheries (MAFF). "Long-Term Prospects for the Demand and Production of Agricultural Products and the Introductory Part of the Basic Direction of Agricultural Policy in the 1980's." Japan's Agricultural Review 80 (December, 1980).

__________________. "On the Implementation of the Basic Direction of the Agricultural Policy in the 1980's. Report by the Agricultural Policy Council." Japan's Agricultural Review. 7, 8(March, 1983).

__________________. Livestock Industry Bureau. "Basic Policies to Promote the Modernization of Dairy Farming and Beef Cattle Raising." Tokyo, October 26, 1983.

________________. *Ammonia Treatment of Straw.* High Mowthorpe Exp. Hus. Farm Leaflet. Tokyo, 1984.

________________. *The Meat Statistics in Japan.* Tokyo, semiannual.

________________. *Abstract of Statistics on Agriculture, Forestry and Fisheries.* Tokyo, annual.

________________. *Statistical Yearbook of the Ministry of Agriculture, Forestry and Fisheries.* Tokyo, annual.

________________. *Monthly Statistics of Agriculture, Forestry and Fisheries.* Tokyo, monthly.

Markert, Clement L. "Cloning Mammals: Current Reality and Future Prospects." *Theriogenology* 21(January, 1984):60-67.

McElroy, Robert G. and Kenneth R. Krause. *New Technologies to Raise Agricultural Efficiencies.* Washington, D.C.: USDA ERS Agr. Info. Bull. 453, August, 1982.

McGrann, James, John Ball, Michael D. Boehlje, and Gene Rouse. *Beef Cattle Feeding in Iowa 1979: Evaluation of Feedlot Systems.* Iowa Coop. Ext. Serv. Pm-602, revised, Iowa State University, July, 1979.

McNeil, Douglas W., Clark R. Busbee and Howard R. Wright, II. "Supply Response to Technological Change and Regulation: The Case of Mechanically Deboned Chicken." *Sou. J. Agr. Econ.* 15(1983): 133-137.

Milligan, Robert, Caroline J. Nowak and Wayne A. Knoblauch. "Profitability of Feeding Dairy Steers to Feeder and Slaughter Weight on Northeast Dairy Farms." Agricultural Economics Staff Paper 81-16, Cornell University, June, 1981.

Ministry of Finance. Monthly Statistics of Foreign Trade. Tokyo, monthly.

Monniaux, D., D. Chupin and J. Saumande. "Superovulatory Responses of Cattle." Theriogenology 19 (January, 1983):55-81.

Mori, Hiroshi and William D. Gorman. "Issues, Facts and Opportunities for Exports of U.S. Beef to Japan." Agribusiness 1(1985):211-218.

Nestel, Barry, ed. Development of Animal Production Systems. New York: Elsevier, 1984.

Organization for Economic Cooperation and Development (OECD). Animal Feeding and Production: New Technical and Economic Developments. Paris, 1981.

Ogura, Takekazu. Can Japanese Agriculture Survive?--A Historical Approach. Tokyo, Agricultural Policy Research Center, 1979.

Ohkawa, Kazushi and Miyohei Shinohara, with Larry Meissner, eds. Patterns of Japanese Economic Development: A Quantitative Appraisal. New Haven: Yale University Press, 1979.

Okimoto, Daniel I., ed. Japan's Economy: Coping with Change in the International Environment. Boulder, Colorado: Westview Press, 1982.

Ota, Yasuji. "What's the Beef About? Perspectives on the Japanese Market." Speaking of Japan, March, 1983, pp. 22-27.

Paarlberg, Philip and Jerry A. Sharples. Japanese and European Community Agricultural Trade Policies: Some U.S. Strategies. Washington, D.C.: USDA FAER 204, August, 1984.

Peacock, F.M. et al. Additive Genetic and Heterosis Effects for Calf and Maternal Influence on Post-Weaning Growth and Carcass Traits of Angus, Brahman and Charolais Crosses. Florida Agr. Exp. Sta. Bull. No. 832 (Technical), University of Florida, 1982.

Pherson, Carl L., Paul R. Hasbargen and Truman R. Nodland. Beef Housing Economics for Farm-Feedlots. Dept. of Agr. and Appl. Econ. Rpt. ER 77-1, University of Minnesota, January, 1977.

Sasaki, Y., H. Iwaisaki, T. Masuno and S. Asoh. "Interaction of Sire X Length of Testing Period and Estimation of Genetic Parameters for Performance Testing Traits of Japanese Black Bulls." J. An. Sci. 55(1982):771-780.

Simpson, James R., L.B. Baldwin, and F.S. Baker, Jr. Investment and Operating Costs for Two Types and Three Sizes of Florida Feedlots. Florida Agr. Exp. Sta. Bull. No. 817, University of Florida, 1981.

Simpson, James R. and Donald E. Farris. The World's Beef Business. Ames: Iowa State University Press, 1982.

Simpson, James R. "Beef Prices and Grading Systems: A World-Wide Perspective." World Animal Review 50(April-June, 1984):29-35.

Simpson, James R. "Beef Production in Japan: Just What is its Potential?" Feedstuffs, July 16, 1984, pp. 15-21.

Simpson, James R. "Projected Impacts of Technological Change on the EEC's Livestock Industry." Feedstuffs, January 14, 1985.

Spedding, C.R.W. Fream's Agriculture, 17th ed. London: John Murray, 1983: 471-493.

Time. "Japan: A Nation in Search of Itself." August 1, 1983, pp. 19-88.

Togamura, Y., A. Miyazaki and R. Kawashima. "Improvement of Nutritive Value of Rice Straw and Hull by Explosion Treatment." Proceedings 5th World Conference on Animal Production, Kyoto, 1984, Vol. 2: 587-588.

United Nations, Food and Agriculture Organization (FAO). Production Yearbook. Rome, various issues.

____________________. Secretariat, Pop. Div., Dept. of Econ. and Soc. App. World Population Prospects, Estimates and Projections as Assessed in 1982. Rome, 1984.

U.S. Department of Agriculture. Economic Indicators of the Farm Sector: Costs of Production, 1982. Washington, D.C.: ECIFS 2-3, 1983.

____________________. Economic Indicators of the Farm Sector: Costs of Production, 1983. Washington, D.C.: ECIFS 2-4, 1984.

____________________. Foreign Agricultural Trade of the United States. Washington, D.C., February, 1985.

____________________. Livestock and Meat Statistics. Washington, D.C., annual.

U.S. Meat Export Federation. The Japanese Beef Market: A Study of Opportunities for MEF Member Firms. Denver, Colorado, December, 1984.

____________________. "A White Paper on U.S. Meat Exports." Denver, Colorado, March, 1985.

Van Arsdall, Roy N. and Kenneth E. Nelson. Characteristics of Farmer Cattle Feeding. Washington, D.C.: USDA AER 503, August, 1983.

Japanese Bibliography

Hokkaido Nōmubu Chikusanka. Nikuyōgyu Kankei Shiryo (Statistics on Beef Cattle). Sapporo, 1984.

Japan Meat Grading Association. Gyu Buta Edaniku Torihiki Kikaku no Kaisetsusho. (Beef and Pork Carcass Trading Standards and Introduction by Color Photography). Tokyo, 1979.

Japan Beef Cattle Association. Nihon Nikuyōgyu Hensenshi (Historical Development of Beef Cattle in Japan). Tokyo, March, 1978.

Kai, S. Nikugyu no Seisan to Ryutsū (Production and Marketing of Beef). Tokyo: Meibun Shobo, December, 1982.

Kajii, I., ed. Chikusan Keiei to Tochiriyō (Livestock Farm Management and Land Utilization). Tokyo: Nosangyoson Bunka Kyokai, December, 1982.

Kishimoto, Y. Gyūniku Keizai Ron (Economic Analysis of the Beef Industry). Tokyo: Chūo Chikusankai, January, 1982.

Livestock Industry Promotion Corporation, ed. Gyuniku no Rekishi (History of Beef in Japan). Tokyo, September, 1978.

Ministry of Agriculture, Forestry and Fisheries (MAFF). Chikusanbutsu Seisanhi Hōkoku (Production Cost of Livestock Products). Tokyo, annual.

__________. Rakuno oyobi Nikuyōgyu Seisan no Kindaika o hakarutameno Kihon Hoshin (Official Guide for Future Dairy and Beef Production). Tokyo, 1983.

__________. Shiryo Geppo. (Monthly Feed Statistics). Tokyo, monthly.

__________. Shokuryō Jyukyū Hyō (Food Balance Sheet Statistics). Tokyo, annual.

Miyazaki, Akira. "Shorai no Gyuniku Seisan" ("Future of Beef Production in Japan"). Nogyo Shinyō Hoken Kyokai. Tokyo, 1979.

Miyazaki, H. Nihongata Chikusan no Shinhōkō (New Direction of Japanese-type Livestock Farming). Tokyo: Ienohikari Kyokai, March, 1984.

Monma, T. Gyuniku no Jyukyudōkō to Shijōtaiō (Demand-Supply Structure of Beef and Its Marketing Strategy). Tokyo: Meibun Shobō, April, 1984.

Mori, H. Gyuniku Mondai to Nihonteki Fūdo (Beef Problems and Japanese Environments). Tokyo: Hyoronsha, December, 1979.

Naito, M. Chikusan Daijiten (Encyclopedia of Animal Production). Tokyo: Yōkendō Press, 1978.

Nosei Chosa Iinkai, ed. Chikusanbutsu no Jyuyōdōkō Bunseki (Demand and Supply Analysis of Livestock Products). Tokyo: Agricultural Policy Research Committee, March, 1984.

Noseikenkyu Center, Shokuryo Hakusho. Gyuniku no Keizai (Economy of Beef). Tokyo, 1980: 59-60.

Shimazu, T. and K. Ozawa. Chikusan Keiei Gaku (Livestock Farm Management). Tokyo: Buneidō, December, 1984.

Shokuniku Torihiki Kitaku Kento Kai. (Meat Trade Grade Discussion Group). Hōkokusho (Report), January, 1985.

Takahashi, I. Gyuniku no Keizaigaku (Economics of Beef). Tokyo: Ochanomizu Shobo, May, 1978.

Takahashi, I. Shokuniku Keizai (Beef Economics: An Analysis of Competitive Structure). Tokyo: Nihon Hyoronsha, May, 1972.

Yokota, T. Gyuniku wa Naze Nihon de Takai no ka. (Why is Beef Expensive in Japan?). Tokyo: Simul Press, Inc., 1977.

Tsuchiya, K., ed. Chikusan Kaihatsuron (Livestock Development). Tokyo: Ochanomizu Shobō, January, 1981.

Yoshida, T. and A. Miyazaki. Amerika no Gyuniku Seisan (Beef Production in the U.S.). Tokyo: Nōrin Tōkei Kyokai, October, 1982.

Zenkoku Nikuyogyu Kyokai. "Nikuyōgyu Shinkō ga Naze Hitsuyōka" ("Why Promotion of the Beef Industry is Requested in Japan"). Tokyo, 1983.

Appendixes

Appendix 1.1. Comparison of land use in selected countries, 1982

Region or Country	Total area	Land area	Arable Land: Arable land and permanent crops	Arable Land: Permanent meadows and pastures	Forest and woodland	Other land
	1,000 ha					
World	13,392,215	13,077,420	1,472,502	3,162,370	4,090,040	4,352,544
Japan	37,231	37,103	4,829	574	25,198	6,479
France	54,703	54,563	18,617	12,775	14,620	8,551
Germany, Fed. Rep.	24,858	24,428	7,462	4,675	7,328	4,963
UK	24,482	24,160	6,978	11,303	2,142	3,737
Canada	997,614	922,097	46,100	24,000	326,129	525,788
USA	936,312	912,680	188,755	237,539	284,464	200,053
Brazil	851,197	845,651	74,670	163,000	570,140	37,841
China	959,696	932,641	100,891	285,690	128,225	417,835
India	328,759	297,319	169,540	11,600	67,520	48,659
Korea	9,848	9,819	2,180	52	6,560	1,027
Australia	768,685	761,793	46,544	444,228	106,000	165,021
USSR	2,240,220	2,227,200	232,266	373,200	920,000	701,734
	Percent[a]					
World		97.6	11.2	24.3	31.3	33.2
Japan		99.7	13.1	1.6	67.9	17.4
France		99.7	34.2	23.5	26.7	15.6
Germany, Fed. Rep.		98.3	30.6	19.3	30.0	20.1
UK		98.7	28.9	46.9	8.8	15.4
Canada		92.4	5.0	2.6	35.4	57.0
USA		97.5	20.7	26.0	31.2	21.9
Brazil		99.3	7.3	19.2	67.7	4.5
China		97.2	10.8	30.6	13.3	45.3
India		90.4	57.0	4.0	22.7	16.3
Korea		99.7	22.3	0.5	66.9	10.3
Australia		99.1	5.7	59.8	13.9	20.6
USSR		99.4	10.4	16.8	41.3	31.5

Source: FAO. Production Yearbook. Rome, 1983.

[a]Arable land, forest and other land are percentages of land area.

Appendix 1.2. Total population, agricultural population, and calculated population density for selected countries, 1983

Region or County	Population			Population density	
	Total	Agricultural	Agricultural population as a proportion of total	Total land	Arable land and permanent crops
	----- 1000 ----		- Percent -	- People per sq. km -	
World	4,669,685	2,075,884	44.5	36	318
Japan	119,260	10,717	9.0	322	2,458
France	54,438	4,070	7.5	100	292
Germany, Fed. Rep.	61,400	2,041	3.3	251	821
UK	56,344	1,027	1.8	233	807
Canada	24,907	1,081	4.3	3	54
USA	234,170	4,404	1.9	26	124
Brazil	131,142	47,014	35.8	16	213
China	1,033,699	591,220	57.2	111	1,024
India	725,456	442,078	60.9	244	428
Korea	40,445	14,154	35.0	412	1,849
Australia	15,003	782	5.2	2	35
USSR	272,300	39,578	14.5	12	118

Source: FAO. Production Yearbook. Rome, 1983 and calculations from Appendix 1.1.

Appendix 1.3. Total population and agricultural population, Japan, 1961-84

Year	Total	Agricultural	Proportion
	- - - - 1,000 - - - -		- Percent -
		FAO data source	
1961	98,275	30,083	30.6
1970	104,345	20,173	19.3
1975	111,524	16,244	14.6
1980	116,782	12,631	10.8
1983	119,260	10,717	9.0
		Japanese data source	
1970	103,720	26,595	25.4
1975	111,940	23,197	20.7
1980	117,060	21,366	18.3
1983	118,602	20,808	17.5
1984	119,316	20,495	17.2

Source: FAO, Production Yearbook, Rome, 1983 and MAFF Abstract of Statistics on Agriculture, Forestry and Fisheries, 1984. Tokyo, March, 1985. The Japanese percent is apparently high because it includes all family members of a family which has agricultural operations. Many of the members will work in non-agricultural jobs. It also includes part-time farm families living in urban areas while FAO data is for farmers and their families who are dependent on agricultural operations.

Appendix 1.4. Number of farm households by size of cultivated land area in Japan, 1970-1984

		Prefectures excluding Hokkaido[a]				
		Farm Size (ha)				
Year	Total farm households	Under 0.5	0.5-1.0	1.0-1.5	1.5-2.0	2.0 and over
		1,000				
1970	5,236	2,030	1,619	874	407	307
1975	4,819	1,195	1,436	727	349	311
1980	4,542	1,911	1,304	652	328	335
1981	4,497	1,848	1,309	661	327	351
1982	4,451	1,828	1,284	648	329	362
1983	4,407	1,809	1,268	640	326	364
1984	4,360	1,792	1,251	629	324	365

		Hokkaido				
		Farm Size (ha)				
Year	Total farm households	Under 2.0	2.0-5.0	5.0-10.0	10.0-20.0	20.0 and over
		1,000				
1970	166	50	50	42	20	5
1975	134	39	37	33	16	10
1980	120	31	29	30	16	11
1981	118	30	27	31	17	14
1982	116	29	26	30	17	15
1983	115	28	26	29	16	15
1984	113	27	26	28	16	16

Source: MAFF. Abstract of Statistics on Agriculture, Forestry and Fisheries, 1984. Tokyo, March, 1985.

[a]Includes Okinawa in 1970 data for comparative purposes even though Okinawa was not part of Japan at that time.

Appendix 1.5. Proportion of households by size of cultivated land, Japan, 1970-1984

Prefectures excluding Hokkaido

Year	Farm Size (ha) Under 0.5	0.5-1.0	1.0-1.5	1.5-2.0	2.0 and over
	- - - - - - - - - - - -	- - - - -	Percent	- - - - -	- - - - - - - - - - -
1970	38.8	30.9	16.7	7.8	5.8
1975	4.14	29.8	15.1	7.2	6.5
1980	42.2	28.7	14.4	7.2	7.4
1981	41.1	29.1	14.7	7.3	7.8
1982	41.1	28.9	14.6	7.4	8.1
1983	41.1	28.8	14.5	7.4	8.3
1984	41.0	28.7	14.4	7.4	8.4

Hokkaido

Year	Farm Size (ha) Under 2.0	2.0-5.0	5.0-10.0	10.0-20.0	20.0 and over
	- - - - - - - - - - - -	- - - - -	1,000	- - - - -	- - - - - - - - - - -
1970	30.1	30.1	25.3	12.0	3.0
1975	29.1	27.6	24.6	11.9	7.5
1980	25.8	24.2	25.0	13.3	9.2
1981	25.4	22.9	26.3	14.4	11.9
1982	25.0	22.4	25.9	14.7	12.9
1983	24.3	22.6	25.2	13.9	13.0
1984	23.9	23.0	24.8	14.2	14.2

Source: Derived from Appendix 1.4.

Note: Percentages may not add to 100 due to rounding errors.

Appendix 1.6. Agricultural labor use by farm household, Japan, 1970-1983

Year	Farm population	Only in own farming	Members per family engaged in farming: Total	Members per family engaged in farming: Engaged in self-farming only	Members per family engaged in farming: Engaged in self-farming with a side job	Labor hours for self-farming: Total	Labor hours for self-farming: Family members	Labor hours for self-farming: Hired workers	Labor hours for self-farming: Hired as a proportion of total
	- - - - 1,000 - - -		- - - - - - - - - Number	- - - - - - -		- - - - - - Hours	- - - - -		- Percent -
1970	26,595	8,519	1.63	1.44	0.10	2,661	2,575	86	03
1975	23,195	6,566	1.41	1.23	0.08	2,245	2,189	56	02
1980	21,366	6,036	1.18	1.07	0.10	1,949	1,874	43	02
1981	21,399	5,766	1.19	1.07	0.10	1,963	1,888	42	02
1982	21,089	5,764	1.17	1.05	0.09	1,941	1,867	43	02
1983	20,808	5,668	1.15	1.03	0.09	1,932	1,885	47	02

Source: MAFF. Abstract of Statistics on Agriculture, Forestry and Fisheries, 1984. Tokyo, March, 1985.

Appendix 1.7. Index numbers of agricultural, forestry and fishery production

Group	1965	1970	1975	1979	1980	1981	1982	1983
				(1980 = 100)				
Agriculture, forestry and fisheries, total	85.8	93.5	99.3	104.7	100.0	100.3	102.1	103.3
Yearly changes (percent)	1.4	1.4	2.5	0.9	4.5	0.3	1.8	1.2
Agriculture, total	85.4	95.9	101.3	107.3	100.0	101.6	103.7	104.3
Field crops, total	103.8	107.9	112.6	111.9	100.0	102.9	104.7	104.8
Rice	125.2	128.1	132.6	121.4	100.0	106.0	105.4	107.6
Wheat and barley	227.3	98.7	47.3	98.6	100.0	99.5	117.7	111.3
Pulses	201.6	165.5	120.0	125.3	100.0	111.2	136.0	110.9
Potatoes and sweet potatoes	159.1	114.7	94.5	100.3	100.0	99.9	109.2	107.9
Vegetables	76.4	90.9	94.0	102.2	100.0	101.8	102.4	99.9
Fruits and nuts	57.1	82.5	101.3	106.4	100.0	94.5	102.2	105.5
Crops for industrial use	110.4	97.3	100.6	104.2	100.0	97.9	99.9	98.2
Sericulture, total	210.2	152.4	124.5	111.2	100.0	88.9	86.9	83.5
Livestock, total	48.2	70.8	79.7	98.6	100.0	99.1	102.1	104.2
Dairy cattle	72.3	102.0	86.3	99.7	100.0	98.3	96.2	96.7
Pigs	37.2	59.3	71.2	97.9	100.0	93.5	96.4	96.0
Layers	77.8	106.7	96.2	101.3	100.0	106.6	110.3	112.3
Broilers	19.8	52.0	72.3	98.3	100.0	99.5	103.1	108.1
Hen eggs	58.8	87.7	89.6	99.5	100.0	99.9	102.7	104.0
Cow milk	49.6	73.3	76.6	99.6	100.0	101.7	104.1	108.7
Forestry, total	141.4	119.7	96.3	98.9	100.0	94.2	94.6	94.3
Logs, total	152.0	129.5	99.3	99.5	100.0	93.1	94.2	93.9
Minor forest products	30.1	49.3	75.7	95.7	100.0	100.9	97.2	98.1
Fisheries, total	62.9	72.0	92.0	96.0	100.0	98.1	99.0	103.5
Marine fisheries	71.5	81.9	97.4	96.2	100.0	96.9	98.7	101.4
Marine culture	25.7	35.6	68.6	93.4	100.0	103.5	99.6	115.2
Whaling	1,403.2	845.0	454.5	96.8	100.0	86.5	88.4	83.7
Inland water fisheries	70.3	76.1	95.3	101.8	100.0	101.1	99.0	99.5
Inland water culture	37.7	49.3	65.9	100.4	100.0	97.5	103.6	100.5

Source: MAFF. Abstract of Statistics on Agriculture, Forestry and Fisheries Production, 1983 and 1984. Tokyo, March, 1984 and March, 1985.

Appendix 1.8. Total value of all exports and imports, Japan, 1965-1984

Year	Yen basis		U.S. dollar basis[a]		
	Exports	Imports	Exports	Imports	Trade balance
	- - Billion Yen - -		- - - - Million dollars - - - -		
1965	3,043	2,941	8,452	8,169	283
1970	6,954	6,797	19,318	18,881	437
1971	8,394	6,910	24,019	19,712	4,307
1972	8,806	7,229	28,591	23,471	5,120
1973	10,031	10,404	36,930	38,314	-1,384
1974	16,208	18,076	55,536	62,110	-6,575
1975	16,545	17,170	55,753	57,863	-2,110
1976	19,935	19,229	67,225	64,799	2,427
1977	21,648	19,132	80,495	70,809	9,686
1978	20,556	16,728	97,543	79,343	18,200
1979	22,532	26,245	103,032	110,672	-7,641
1980	29,382	31,995	129,807	140,528	-10,721
1981	33,469	31,464	152,630	143,290	8,741
1982	34,433	32,656	138,831	131,931	6,900
1983	34,909	30,015	146,927	126,393	20,554
1984	40,325	32,321	168,270	123,920	44,350

Source: Adapted from MAFF. Abstract of Statistics on Agriculture, Forestry, and Fisheries, 1983. Tokyo, March, 1984 and MAFF, Monthly Statistics of Agriculture, Forestry and Fisheries (selected issues).

[a]Fixed exchange rate was ¥ 360/U.S. $1.00 until September 1971. After that a floating rate has prevailed. See Appendix 1.9 for rates.

Appendix 1.9. Exchange rate, Japanese yen per United States dollar, 1960-1984

Japan fiscal year	Exchange rate
	- Yen per dollar -
1960-1970	360
1971	351
1972	303
1973	272
1974	292
1975	297
1976	297
1977	269
1978	210
1979	219
1980	227
1981	221
1982	249
1983	238
1984	240

Source: Ministry of Finance, *Japan Exports and Imports*, annual issues.

Appendix 1.10. Value of foreign trade of agricultural, forest and fishery products, Japan 1970–1984

Year	Exports: Total, agricultural and non-agricultural	Exports: Agricultural forest and fishery products[a]	Exports: Agricultural products only[a]	Exports: Agricultural products as proportion of total	Imports: Total agricultural and non-agricultural	Imports: Agricultural, forest and fishery products[a]	Imports: Agricultural products only[a]	Imports: Agricultural products as proportion of total
	- - - - Million Yen	- - - -	- - - -	- Percent -	- - - - Million Yen	- - - -	- - - -	- Percent -
1970	6,954,367	315,767	139,718	2.0	6,797,221	2,249,492	1,511,276	22.2
1975	16,545,314	308,364	115,038	0.7	17,170,027	4,639,411	3,325,545	19.4
1977	21,648,070	349,904	111,152	0.5	19,131,780	5,213,844	3,367,126	17.6
1978	20,555,841	333,400	112,254	0.5	16,727,624	4,498,351	2,821,793	16.9
1979	22,531,539	395,963	165,013	0.7	24,245,351	6,318,655	3,604,940	14.9
1980	29,382,472	506,174	208,912	0.7	31,995,325	6,619,308	4,006,575	12.5
1981	33,468,985	522,476	238,937	0.7	31,464,146	6,111,630	4,052,702	12.9
1982	34,432,501	487,001	197,531	0.6	32,656,303	6,389,801	4,019,413	12.3
1983	34,909,269	497,288	201,736	0.6	30,014,784	6,062,627	3,065,504	10.2
1984	40,325,294	532,754	202,285	0.5	32,321,127	6,575,995	4,407,800	13.6

Source: MAFF. Abstract on Agriculture, Forestry and Fisheries, 1984. Tokyo, March, 1985 and MAFF Monthly Statistics of Agriculture, Forestry and Fisheries, selected issues.

[a]Includes livestock products and raw silk.

Appendix 1.11. Value of imported field crop products, Japan, 1970–1984

Year	Total	Cereals, their flour and preparations	Fruits & nuts and their preparations	Vegetables and their preparations	Sugar	Tobacco	Cotton
	- - - - - - - - - -	- - - - - - - - - -	- - - - - - - - - -	Million Yen	- - - - - - - - - -	- - - - - - - - - -	- - - - - - - - - -
1970	1,193,399	376,614	85,154	21,286	122,162	26,579	169,463
1975	2,770,999	966,214	125,525	51,193	527,920	94,966	251,164
1977	2,663,699	758,098	149,525	89,419	218,327	89,303	314,003
1978	2,129,174	641,201	149,917	81,887	144,750	91,313	227,440
1979	2,700,440	808,147	187,170	95,076	186,210	86,646	273,768
1980	3,180,014	998,691	189,386	107,757	310,854	84,630	311,031
1981	3,135,456	1,089,115	210,664	140,377	191,354	106,943	309,035
1982	3,050,720	975,066	238,117	138,659	143,059	125,189	313,387
1983	3,065,504	990,943	231,337	130,894	117,643	125,917	272,089
1984	3,402,791	1,118,126	249,498	161,878	95,521	126,477	316,407

Source: MAFF. Abstract of Statistics on Agriculture, Forestry and Fisheries, 1984. Tokyo, March, 1985.

Appendix 1.12. Value of imported livestock products, total products, and total fishery products, Japan, 1970–1984

Year	Livestock and its products				Animal oils and fats	Hides, skins, wool	Other livestock products	Total forest products	Total fishery products
	Total	Live animals	Meat and its preparations	Dairy products and eggs					
	Million Yen								
1970	287,531	7,382	47,521	23,575	21,592	169,420	18,041	623,588	114,628
1975	522,174	12,664	190,420	50,284	20,816	219,267	28,721	928,337	385,529
1979	852,797	21,718	358,432	65,903	20,480	334,190	52,074	1,782,955	930,760
1980	790,514	18,880	331,604	77,909	20,719	288,929	52,473	1,848,414	764,319
1981	903,000	17,891	411,928	82,351	15,160	326,250	49,420	1,179,046	879,881
1982	936,803	14,104	413,381	99,123	13,559	346,322	50,314	1,323,643	1,046,745
1983	862,804	15,731	407,791	78,113	13,061	294,559	54,550	1,099,511	1,003,124
1984	980,334	16,255	436,702	71,528	13,371	382,543	59,937	1,117,975	1,050,220

Source: MAFF. Abstract of Statistics on Agriculture, Forestry and Fisheries, 1984. Tokyo, March, 1985.

Appendix 1.13. Beef production, trade and consumption in Japan, 1960–1983

Japan fiscal year	Domestic production	Imports	Exports	Change in stocks	Supplies for domestic consumption	Domestic utilization: Waste	Domestic utilization: Gross Food	Domestic utilization: Ratio	Domestic utilization: Net food	Annual per capita consumption
	- - - 1,000 Metric Tons - - -							- Percent -	1,000 MT -	- Kg -
1960	141	6	—	—	147	3	144	77	111	1.2
1961	141	6	—	—	147	3	144	77	111	1.2
1962	153	4	—	—	157	3	154	77	119	1.3
1963	198	5	—	—	203	4	199	77	153	1.6
1964	229	6	—	—	235	5	230	77	177	1.8
1965	190	11	—	—	201	4	197	77	152	1.5
1966	152	14	—	—	166	3	163	77	126	1.3
1967	148	16	—	—	164	3	161	77	124	1.2
1968	172	15	—	—	187	4	183	77	141	1.4
1969	230	18	—	—	248	5	243	77	187	1.8
1970	265	33	—	—	298	6	292	72	210	2.0
1971	279	62	—	—	341	7	334	72	240	2.3
1972	290	73	—	—	363	7	356	72	256	2.4
1973	215	170	—	28	357	7	350	72	252	2.3
1974	326	40	—	-18	384	8	376	72	271	2.5
1975	327	91	—	11	407	8	399	70	279	2.5
1976	309	134	—	-7	450	9	441	70	309	2.7
1977	371	132	—	6	497	10	487	70	341	3.0
1978	406	146	—	-3	555	11	544	70	381	3.3
1979	400	189	—	5	584	12	572	70	400	3.4
1980	431	172	—	6	597	12	585	70	410	3.5
1981	476	172	—	16	632	13	619	70	433	3.7
1982	483	198	—	—	681	14	667	70	467	3.9
1983	505	208	—	-11	724	14	710	70	497	4.2

Source: MAFF. Food Balance Sheet. Tokyo, various issues.

Note: Per capita consumption is retail weight basis.

Appendix 1.14. Pork production, trade and consumption in Japan, 1960–1983

Japan fiscal year	Domestic production	Imports	Exports	Change in stocks	Supplies for domestic consumption	Domestic utilization: Waste	Domestic utilization: Gross Food	Domestic utilization: Ratio	Domestic utilization: Net food	Annual per capita consumption
	- - 1,000 Metric Tons - -							- Percent -	- 1,000 MT -	- Kg -
1960	149	6	—	—	155	3	152	82	125	1.3
1961	240	1	—	—	241	5	236	82	194	2.1
1962	322	—	—	—	322	7	315	82	258	2.7
1963	271	8	—	—	279	6	273	82	224	2.3
1964	314	2	—	—	316	6	310	82	254	2.6
1965	385	—	—	—	385	8	377	82	309	3.1
1966	542	—	—	32	510	10	500	82	410	4.1
1967	546	—	—	-23	569	11	558	82	458	4.6
1968	511	18	—	-11	540	11	529	82	434	4.3
1969	525	36	—	—	561	11	550	82	451	4.4
1970	691	17	—	—	708	14	694	70	486	4.7
1971	755	29	—	—	784	16	768	70	538	5.1
1972	780	89	—	—	869	17	852	70	596	5.6
1973	884	128	—	—	1,012	20	992	70	694	6.4
1974	966	71	—	—	1,037	21	1,016	70	711	6.5
1975	891	208	—	41	1,058	21	1,037	70	726	6.5
1976	1,096	187	—	15	1,268	25	1,243	70	870	7.7
1977	1,189	161	—	-23	1,373	27	1,346	70	942	8.3
1978	1,323	155	—	10	1,468	29	1,439	70	1,007	8.7
1979	1,465	176	—	15	1,626	33	1,593	70	1,115	9.6
1980	1,430	207	—	-9	1,646	33	1,613	70	1,129	9.6
1981	1,409	232	—	-1	1,642	33	1,609	70	1,126	9.6
1982	1,427	199	—	-21	1,647	33	1,614	70	1,130	9.5
1983	1,433	271	—	23	1,681	34	1,647	70	1,153	9.6

Source: MAFF. Food Balance Sheet. Tokyo, various issues.

Note: Per capita consumption is retail weight basis.

Appendix 1.15. Chicken meat production, trade and consumption in Japan, 1960-1983

Japan fiscal year	Domestic production	Imports	Exports	Change in stocks	Supplies for domestic consumption	Domestic utilization				Annual per capita consumption
						Waste	Gross Food	Ratio	Net food	
	- - - 1,000 Metric Tons - - -							- Percent -	- 1,000 MT -	- Kg -
1960	44	—	—	—	44	1	43	77	33	.2
1961	97	—	—	—	97	2	95	77	73	.8
1962	121	—	—	—	121	2	119	77	92	1.0
1963	142	5	—	—	147	3	144	77	111	1.2
1964	181	4	—	—	185	4	181	77	139	1.4
1965	205	8	—	—	213	4	209	77	161	1.6
1966	261	7	—	—	268	5	263	77	203	2.0
1967	303	10	—	—	313	6	307	77	236	2.4
1968	336	18	—	—	354	7	347	77	267	2.6
1969	423	20	—	—	443	9	434	77	334	3.3
1970	505	12	—	—	517	10	507	77	390	3.8
1971	571	30	1	—	600	12	588	77	453	4.3
1972	638	27	1	—	664	14	650	77	501	4.7
1973	706	26	1	—	731	15	716	77	551	5.1
1974	730	21	—	—	749	15	734	77	565	5.1
1975	756	28	3	—	781	16	765	77	589	5.2
1976	834	40	2	—	872	17	855	77	658	5.8
1977	944	48	3	6	983	20	963	77	742	6.5
1978	1,022	66	3	—	1,085	33	1,068	77	819	7.1
1979	1,095	69	3	-5	1,166	23	1,143	77	880	7.6
1980	1,120	80	4	2	1,194	24	1,170	77	901	7.7
1981	1,140	104	3	3	1,238	25	1,213	77	934	7.9
1982	1,200	107	3	2	1,302	26	1,276	77	983	8.3
1983	1,257	100	2	-4	1,359	27	1,332	77	1,026	8.6

Source: MAFF. Food Balance Sheet. Tokyo, various issues.

Appendix 2.1. Daily per capita caloric intake, Japan, 1930–1980

Year	Grain		Potatoes	Pulses	Vegetables	Fruits	Meats
	Rice	Other	& starch				
	Kilocalories						
1930	1,248	249	94	70	51	25	7
1935	1,185	252	96	71	51	28	7
1940	1,303	234	92	75	51	27	9
1945	872	240	183	12	38	9	3
1950	1,036	491	152	15	44	19	8
1955	1,063	424	166	95	73	17	17
1960	1,106	333	141	104	84	29	28
1965	1,090	332	131	107	74	39	52
1970	928	335	115	117	78	53	81
1975	857	341	110	110	76	57	108
1980	770	353	152	99	76	53	138

Year	Eggs	Milk & dairy products	Marine products	Sugar	Fats & oils	Bean curd and soybean	Total
	Kilocalories						
1930	8	4	63	145	21	60	2,045
1935	10	5	67	143	21	60	1,996
1940	9	6	55	122	33	60	2,076
1945	2	3	36	6	3	42	1,449
1950	3	9	71	34	19	44	1,945
1955	16	20	84	128	67	59	2,217
1960	27	36	87	157	105	53	2,290
1965	50	62	99	196	167	60	2,458
1970	64	82	102	283	239	57	2,533
1975	61	87	119	262	288	51	2,586
1980	64	102	133	245	351	49	2,586

Source: MAFF, *Food Balance Sheet*, various years.

Appendix 3.1. Production, trade and consumption of dairy products in Japan, 1960–1983[a]

Japan fiscal year	Production	Imports	Exports	Change in stocks	Supplies for domestic consumption	Utilization			Annual per capita consumption
						Feed	Waste	Food	
	- - - - - - - - - -	- - - -	- - - -	1,000 Metric Tons	- - - -	- - - -	- - - -	- - - -	- Kg -
1960	1,939	237	0	—	2,176	58	40	2,078	22.2
1961	2,180	276	0	—	2,456	58	46	2,352	24.9
1962	2,526	357	0	60	2,823	67	55	2,701	28.4
1963	2,837	481	0	30	3,288	67	65	3,156	32.8
1964	3,053	486	0	41	3,850	67	68	3,445	35.4
1965	3,271	506	0	38	3,815	57	72	3,686	37.5
1966	3,431	841	0	5	4,277	65	82	4,130	41.7
1967	3,663	964	0	135	4,492	65	87	4,340	43.3
1968	4,140	629	0	39	4,730	69	89	4,572	45.1
1969	4,575	568	0	117	5,026	80	94	4,852	47.3
1970	4,789	561	0	-5	5,355	58	102	5,195	50.1
1971	4,841	569	0	-77	5,487	46	106	5,335	50.7
1972	4,944	746	0	-29	5,719	45	111	5,563	51.7
1973	4,898	1,032	0	27	5,903	44	115	5,744	52.9
1974	4,876	1,038	0	36	5,878	43	112	5,723	51.8
1975	5,008	979	0	-136	6,123	41	117	5,965	53.3
1976	5,369	1,169	0	208	6,330	44	120	6,166	54.5
1977	5,846	1,016	0	178	6,684	50	126	6,508	57.0
1978	6,256	1,045	0	290	7,011	52	132	6,827	59.3
1979	6,464	1,109	0	106	7,467	136	140	7,191	61.9
1980	6,498	1,027	8	-42	7,559	143	141	7,275	62.1
1981	6,612	1,068	5	-241	7,916	125	149	7,642	64.8
1982	6,844	1,085	6	-151	8,074	84	153	7,841	66.0
1983	7,088	1,092	0	-54	8,234	56	158	8,020	67.1

Source: MAFF. Food Balance Sheet. March, 1985.

[a]Fluid basis.

Appendix 3.2. Beef breed inventory in Japan, 1981

Breed	Total inventory[a]	Percent of total	Bulls only
Japanese Black (Wagyu)	710,484	86.8	1,187
Japanese Brown (Wagyu)	72,507	8.9	157
Japanese Poll (Wagyu)	1,871	0.2	11
Japanese Shorthorn	25,137	3.1	485
Angus	3,281	0.4	55
Hereford	4,428	0.5	77
Charolais	66	0.0	5
Others	425	0.0	9
Total	818,199	100.0	1,986

Source: Longworth: Beef in Japan. New York, University of Queensland Press, 1983, p. 86.

[a]Includes all cows, heifers and bulls kept for breeding.

Appendix 3.3. Dairy and other cattle slaughter in Japan, 1965–1984

Year	Cows and heifers	Steers	Other cattle	Subtotal	Calves	Total
	- - - -	- - - -	- - - -	Head - - - -	- - - -	- - - -
1965	228,324	—	—	228,324	238,959	467,283
1966	195,634	—	—	195,634	151,664	347,298
1967	192,901	36,337	—	229,238	145,329	374,567
1968	180,387	98,000	—	278,387	172,766	451,153
1969	222,160	153,677	—	375,837	318,196	694,033
1970	315,117	130,686	—	445,803	308,708	754,511
1971	378,217	107,122	—	485,339	221,480	706,819
1972	379,354	182,535	—	561,889	121,082	682,971
1973	299,053	188,681	—	487,734	33,026	520,760
1974	355,061	307,844	—	662,905	169,418	832,323
1975	395,493	341,475	—	736,968	124,849	861,817
1976	316,146	230,459	—	546,605	74,288	620,893
1977	348,357	318,358	—	666,715	89,508	756,223
1978	353,128	353,076	—	706,204	63,215	769,419
1979	392,025	361,390	—	753,415	41,242	794,657
1980	416,946	395,946	—	812,892	42,926	855,818
1981	486,286	466,639	—	952,925	42,490	995,415
1982	486,408	470,991	10,272	967,671	41,523	1,009,194
1983	473,412	459,802	9,871	943,085	42,566	985,651
1984	480,294	458,229	9,851	948,374	41,019	989,393

·Source: MAFF. The Meat Statistics in Japan, Tokyo, January 1, 1985 and Monthly Statistics, March, 1985.

Appendix 3.4. Wagyu cattle slaughter in Japan, 1965-1984

Year	Females	Steers	Bulls	Subtotal	Calves	Total
	- - - - - - - - - -	- - - - -	- - - - - Head	- - - - -	- - - - -	- - - - -
1965	311,042	251,717	124,810	687,569	13,558	701,127
1966	234,208	189,995	35,472	459,675	5,994	465,669
1967	176,492	161,182	35,365	373,035	5,067	378,102
1968	152,038	164,493	33,662	350,093	4,636	354,729
1969	210,865	210,138	33,413	454,416	5,128	459,544
1970	290,475	221,063	28,674	540,212	3,808	544,020
1971	299,562	220,497	24,431	544,490	2,682	547,172
1972	282,356	219,391	23,204	524,951	2,693	527,644
1973	158,982	142,788	16,397	318,167	2,606	320,773
1974	135,801	169,170	15,424	320,395	1,936	322,331
1975	179,807	212,807	13,507	406,121	2,272	408,393
1976	196,319	191,967	12,312	400,598	2,095	402,693
1977	224,078	203,129	12,656	439,863	1,826	441,689
1978	245,555	236,604	9,771	491,930	1,782	493,712
1979	208,278	217,326	6,915	432,519	1,471	433,990
1980	166,413	202,735	4,999	374,147	1,318	375,465
1981	150,856	213,544	4,364	368,764	1,543	370,307
1982	163,372	220,085	4,179	387,636	1,632	389,268
1983	200,680	240,280	4,161	445,121	2,254	447,375
1984	269,900	269,822	4,335	544,057	2,611	546,668

Source: MAFF. The Meat Statistics in Japan, Tokyo, January 1, 1985 and Monthly Statistics, March, 1985.

Appendix 4.1. Components of formula feed, all livestock, Japan, 1965-1983

Japan fiscal year	Corn	Sorghum	Wheat	Barley	Rice	Other grains	Wheat bran	Rice bran	Rice bran oilcake	Other brans	Alfalfa meal pellets	Soybean meal
	1,000 Metric Tons											
1965	2,869	1,544	21	125	--	145	554	74	241	--	423	623
1970	4,417	3,972	140	129	--	467	500	154	358	456	412	1,469
1975	6,263	3,815	15	157	--	399	495	145	321	532	332	1,787
1980	10,359	3,689	145	198	--	198	633	158	282	811	312	2,489
1981	10,369	3,456	161	187	--	410	673	155	290	856	290	2,322
1982	10,537	3,080	149	194	--	1,049	707	174	290	862	276	2,464
1983	11,477	3,291	141	225	--	867	742	191	273	945	315	2,591
	Percent											
1965	35.0	18.9	0.3	1.5	--	1.8	6.8	0.9	2.9	--	5.2	7.6
1970	29.4	26.3	0.9	0.9	--	3.1	3.3	1.0	2.4	3.0	2.7	9.7
1975	37.1	22.6	0.1	0.9	--	2.4	2.9	0.9	1.9	3.1	2.0	10.6
1980	46.2	16.6	0.7	0.9	--	1.3	2.9	0.7	1.3	3.6	1.4	11.2
1981	46.8	15.6	0.7	0.9	--	1.8	3.0	0.7	1.3	3.9	1.3	10.5
1982	46.0	13.5	0.6	0.8	--	4.6	3.1	0.8	1.3	3.7	1.2	10.8
1983	47.1	13.5	0.6	0.8	--	3.6	3.0	0.8	1.1	3.9	1.3	10.6

Source: Japan Feed Association. Shiryo Geppo (Feed Monthly), Tokyo, various issues.

Appendix 4.2. Components of formula feed, beef cattle, Japan, 1965-1983

Japan fiscal year	Corn	Sorghum	Wheat	Barley	Rice	Other grains	Wheat bran	Rice bran	Rice bran oilcake
	1,000 Metric Tons								
1965	5	16	—	13	—	2	15	2	4
1970	75	277	7	58	—	79	50	33	23
1975	295	477	1	95	—	103	92	47	25
1980	852	621	20	137	—	95	165	71	37
1981	903	532	31	131	7	76	175	64	39
1982	941	454	33	135	94	66	201	67	39
1983	931	432	30	147	—	141	205	66	38
	Percent								
1965	06	20	—	16	—	02	18	02	05
1970	09	32	01	07	—	09	06	04	03
1975	19	31	00	06	—	07	06	03	02
1980	31	23	01	05	—	04	06	03	01
1981	34	20	01	05	00	03	07	02	01
1982	34	17	01	05	03	02	07	02	01
1983	35	16	01	06	00	05	08	03	01

Japan fiscal year	Other brans	Alfalfa meal pellets	Other soybean meal	Vege-table oilcakes	Fish meal	Fish solubles	Skim milk powder	Other animal meat	Total
	1,000 Metric Tons								
1965	—	3	7	3	—	—	—	—	82
1970	25	36	66	59	—	—	—	—	861
1975	80	67	69	45	2	—	—	1	1,539
1980	149	113	141	77	3	1	14	1	2,706
1981	163	98	127	84	3	1	13	1	2,648
1982	176	84	131	93	3	1	14	1	2,725
1983	193	84	103	90	—	—	—	—	2,634
	Percent								
1965	—	04	09	04	—	—	—	—	100
1970	03	04	08	07	—	—	—	—	100
1975	05	04	04	03	00	—	—	00	100
1980	06	04	05	03	00	00	00	00	100
1981	06	04	05	03	00	00	00	00	100
1982	06	03	05	03	00	00	00	00	100
1983	07	03	04	03	—	—	—	—	100

Source: Japan Feed Association. Shiryo Geppo (Feed Monthly), Tokyo, various issues.

Appendix 4.3. Coarse grain import composition between animal and human use, Japan, 1983 and 1984

Item	Quantity 1983	Quantity 1984	Proportion 1983	Proportion 1984
	- 1,000 Metric Tons -		- - - Percent - - -	
Corn				
Feed	11,011	10,274	75	73
Other	3,690	3,896	25	27
Total	14,701	14,170	100	100
Sorghum				
Feed	2,740	4,143	93	93
Other	217	335	07	07
Total	2,957	4,478	100	100
Barley				
Feed	1,458	1,530	99	98
Other	19	37	01	02
Total	1,477	1,567	100	100
Oats				
Feed	46	52	62	41
Other	28	74	38	59
Total	74	126	100	100
Rye				
Feed	117	315	88	92
Other	16	27	12	08
Total	133	342	100	100
Millet				
Feed	48	36	100	100
Other	0	0	0	0
Total	48	36	100	100

Source: Ministry of Finance. Japan Exports and Imports. Tokyo, annual issues.

Appendix 4.4. Imports of grains by country of origin, Japan, 1960-1984

Calendar year	United States	Canada	Australia	Argentina	Thailand	South Africa	Other	Total
	- - - - - - - - - - - - - - - 1,000 Metric Tons - - - - - - - - - - - - - - -							
1960	229	2	2	428	314	254	175	1,404
1965	3,859	245	175	206	596	30	517	5,628
1970	6,588	622	513	1,757	562	399	385	10,826
1975	7,381	1,032	1,558	887	818	1,037	413	13,106
1980	15,636	703	1,108	98	230	913	19	18,707
1981	15,296	941	880	58	5	1,444	15	18,639
1982	13,386	942	688	886	206	2,398	—	18,496
1983	14,388	1,187	312	2,253	17	1,030	83	19,270
1984	16,021	1,039	1,659	1,740	10	10	240	20,479
	- - - - - - - - - - - - - - - - Percent - - - - - - - - - - - - - - - -							
1960	16	—	—	30	22	18	12	100
1965	69	03	03	04	11	01	09	100
1970	60	06	05	16	05	04	04	100
1975	56	08	12	07	06	08	03	100
1980	84	04	06	—	01	05	—	100
1981	82	05	05	—	—	08	—	100
1982	72	05	03	05	01	13	—	100
1983	75	06	02	12	—	05	—	100
1984	78	05	08	08	—	—	01	100

Source: Ministry of Finance. Japan Exports and Imports. Tokyo, annual issues.

Appendix 4.5. Imports of oilseeds used in livestock production by Japan, 1961-1984

Calendar year	Soybeans	Rapeseed	Cottonseed	Sesame	Copra	Total
	1,000 Metric Tons					
1961	1,158	24	100	22	80	1,324
1965	1,847	108	217	33	94	2,299
1970	3,244	345	297	53	127	4,066
1975	3,334	668	116	39	90	4,247
1980	4,401	1,067	91	63	65	5,687
1981	4,197	1,169	66	59	74	5,565
1982	4,344	1,204	81	69	82	5,780
1983	4,995	1,201	85	82	66	6,429
1984	4,515	1,299	105	75	73	6,067
	Percent[a]					
1961	87	02	08	02	06	100
1965	80	05	09	01	04	100
1970	80	08	07	01	03	100
1975	79	16	03	01	02	100
1980	77	19	02	01	01	100
1981	75	21	01	01	01	100
1982	75	21	01	01	01	100
1983	78	19	01	01	01	100
1984	74	21	02	01	01	100

Source: Ministry of Finance. Japan Exports and Imports. Tokyo, annual issues and MAFF. Abstract of Statistics on Agricultural, Forestry and Fisheries, 1984. Tokyo, March, 1985.

[a]May not add to 100 due to rounding.

Appendix 4.6. Feed use per unit of livestock products in Japan, 1970-1990[a]

Item	Year				
	1970	1975	1980	1985	1990
	- - - - - - - - - - Kg Feed/Kg Product - - - - - - -				
Beef and veal					
Total concentrates	4.18	5.61	8.08	9.98	11.68
Grains	2.84	3.99	5.72	7.15	8.41
Protein meal	.44	.33	.56	.69	.81
Pork					
Total concentrates	5.36	4.36	4.34	4.63	4.59
Grain	3.35	3.00	3.17	3.47	3.44
Protein meal	.94	.83	.86	.92	.91
Chicken					
Total concentrates	2.07	3.13	2.90	2.91	2.78
Grain	2.02	2.06	1.99	2.03	2.01
Protein meal	.66	.70	.63	.63	.60
Eggs					
Total concentrates	4.00	3.65	3.68	3.42	3.30
Grain	2.63	2.41	2.52	2.33	2.26
Protein meal	.85	.82	.81	.75	.73
Milk					
Total concentrates	.42	.44	.43	.42	.40
Grain	.17	.20	.21	.22	.23
Protein meal	.09	.08	.07	.07	.07

Source: From Coyle, W.T. Japan Feed-Livsetock Economy. FAER No. 177. USDA, Washington D.C. 1983, which is based on Shiryo Geppo (Feed Monthly) and The Meat Statistics in Japan.

[a]The rates are not feed conversions, but rather are indicators of total feed use as they are obtained by dividing total feed by total animals.

Appendix 4.7. Projections of consumption and imports of course grains in Japan, 1985 and 1990

Item	1980	1985 Low	1985 High	1990 Low	1990 High
	- - - - - - - -	- - - - 1,000	Metric Tons - - - -	- - - -	- - - -
Domestic production	400	474	474	577	577
Consumption total	19,197	24,089	25,307	29,659	32,762
Industrial	3,086	3,827	3,827	4,747	4,747
Major livestock feeds	15,751	19,809	21,000	24,355	27,389
Other feeds	360	453	480	557	626
Imports	18,863	23,615	24,833	29,082	32,185

Source: Coyle, W.T. *Japan's Feed-Livestock Economy*. FAER No. 177, USDA Washington, D.C., 1983.

Appendix 4.8. Projections of protein meal demand in Japan, 1990

Item	Livestock production in 1980	Protein meal demand	MAFF 1990 projections[1]		ERS projection I Low		ERS projection II High	
			Livestock production	Protein meal demand	Livestock production	Protein meal demand	Livestock production	Protein meal demand
	1,000 Metric Tons							
Beef and veal	418	234	630	510	529	428	596	483
Pork	1,476	1,296	1,940	1,765	2,498	2,273	2,799	2,547
Chicken	1,145	721	1,460	876	1,868	1,121	2,112	1,267
Eggs	1,999	1,619	2,220	1,621	2,398	1,751	2,695	1,967
Dairy products	6,502	455	8,420	589	9,298	651	10,485	734
Total protein meal utilization in formula feed	—	4,299	—	5,361	—	6,224	—	6,998
Average annual increment, 1980–1990	—	—	—	106	—	193	—	270
	Percent							
Average annual growth rate, 1980–1990	—	—	—	2.2	—	3.8	—	5.0

— = Not applicable.

Source: Coyle, W.T. Japan's Feed-Livestock Economy. FAER No. 177. Washington, D.C., 1983.

[1]Based on MAFF's Long-Term Prospects for the Demand and Supply of Agricultural Products, November 1980.

Appendix 4.9. Japan: Projections of consumption and imports of oilseeds and protein meal in Japan, 1985 and 1990

Item	1980	1985 projections Low	1985 projections High	1990 projections Low	1990 projections High
	------------ 1,000 Metric Tons ------------				
Oilseeds[1]					
Production[2]	259	354	354	525	525
Imports[3]	5,899	6,665	6,977	7,473	8,307
	(4,401)	(5,199)	(5,442)	(5,903)	(6,562)
Total utilization	6,158	7,019	7,331	7,998	8,832
Oilseeds not crushed[4]	1,283	1,215	1,215	1,289	1,289
Oilseeds crushed[5]	4,875	5,804	6,116	6,709	7,543
Protein meal (soybean equivalent)					
Production	4,233	4,907	5,119	5,656	6,240
Vegetable oilmeal (domestic and imported oilseeds)[6]	3,261	3,947	4,159	4,696	5,280
Fishmeal[7]	972	960	960	960	960
Imports[3]	534	758	909	1,150	1,412
	(326)	(470)	(563)	(736)	(904)
Exports	63	--	--	--	--
	(1)				
Apparent utilization	4,704	5,665	6,028	6,806	7,652
Formula feed[8]	4,299	5,198	5,512	6,224	6,998
Other uses[9]	402	467	516	582	654

-- = None or negligible.

Numbers in parentheses show the soybean component of the total amount.

[1]Includes soybeans, rapeseed, cottonseed, sesame, copra, linseed, palm kernel, and peanuts.

[2]Based on MAFF's The Long-Term Prospects for the Demand and Supply of Agricultural Products which projects expanded production and area for soybeans and peanuts.

[3]Oilseed imports are expected to increase more slowly than protein meal imports, reflecting conservative expansionary behavior on the part of crushers.

[4]Oilseeds not crushed are principally soybeans, peanuts, and sesame, which are consumed as food. Soybeans used in traditional Japanese dishes represent about 80 percent of this category. No growth is expected in this category.

[5]The crush rate for oilseeds is based on a weighted average of crush rate estimates for the oilseeds listed in footnote 1. The rate is expected to grow from 79 percent in 1980 to about 85 percent in 1990 because of no growth in the food category.

[6]The meal yield for crushed oilseeds is estimated at about 67 percent in 1980. This is expected to increase to about 70 percent in 1990 with more extensive use of soybeans.

[7]Fishmeal production is expected to show little or no growth in the 1980's because of higher real fish prices and more extensive use of low-grade fish for human consumption.

[8]Computed from data in tables 50 and 51, in Coyle (1983).

[9]Includes protein meal uses in simple feeds and for industrial purposes. Growth in this category is assumed to parallel that protein meal utilization in formula feed.

Source: Coyle W.T. Japan's Feed-Livestock Economy. FAER No. 177. USDA, Washington, D.C., 1983.

Appendix 5.1. Number and grade of Wagyu females slaughtered and graded in Japan, 1972-1984

Year	Total Wagyu graded: All cattle	Females	Proportion females	Grade: Supreme Tokusen	Superior Gokujyu	Excellent Jyo	Medium Chu	Common Nami	Utility Togai
	- - - Head - - -		- Percent -	- - - Percent - - -					
1972	123,647	68,368	55.3	1.7	9.3	38.7	42.2	7.2	0.8
1973	87,827	43,524	49.6	2.0	11.0	42.2	38.9	4.9	0.9
1974	103,237	42,585	41.3	2.8	10.0	36.3	41.8	7.7	1.5
1975	132,288	52,871	40.0	2.2	8.1	26.6	41.7	16.8	4.6
1976	144,251	64,783	44.9	1.8	6.7	23.4	46.3	18.6	3.2
1977	171,183	80,147	46.8	1.4	4.8	20.6	49.0	21.2	3.0
1978	205,302	91,079	44.4	1.1	4.2	19.2	52.4	20.5	2.6
1979	186,405	79,954	42.9	1.3	5.1	22.4	50.1	19.1	2.0
1980	171,039	66,067	38.6	2.0	6.0	24.2	47.3	18.3	2.2
1981	182,467	61,096	33.5	1.6	5.5	20.3	44.8	24.2	3.6
1982	199,678	65,557	32.8	1.3	4.5	17.2	41.1	30.0	5.8
1983	238,854	87,319	36.6	1.0	3.1	13.7	38.7	36.5	7.0
1984	315,466	134,883	42.8	0.8	2.4	10.5	36.4	43.4	6.4

Source: Derived from MAFF. The Meat Statistics of Japan. Tokyo, 1985.

Appendix 5.2. Number and grade of female dairy cattle slaughtered and graded in Japan, 1972-1984

	Total dairy cattle graded			Grade of Females					
Year	Females	Graded	Proportion	Supreme Tokusen	Superior Gokujyu	Excellent Jyo	Medium Chu	Common Nami	Utility Togai
	- - - - Head - - - -		- Percent -	- - - - - - - - - - - - - - Percent - - - - - - - - - - - - - -					
1971	132,585	83,095	62.7	--	0	1.2	18.5	51.5	28.8
1973	114,042	64,456	56.5	0	0.1	3.2	24.1	45.0	27.5
1974	177,625	80,833	45.5	0	0.1	1.8	23.4	42.3	32.5
1975	205,679	101,692	49.4	0	0.1	1.4	17.3	43.3	37.9
1976	176,910	93,976	53.1	0	0.2	3.2	24.2	38.9	33.5
1977	243,549	119,110	48.9	0	0.2	2.8	22.6	41.0	33.4
1978	269,591	124,305	46.1	0	0.2	2.5	23.9	42.2	31.2
1979	302,638	146,406	48.4	0	0.3	.1	22.7	42.8	31.1
1980	341,601	160,295	46.9	0	0.4	3.9	23.9	41.0	30.8
1981	419,322	200,105	47.7	0	0.4	3.5	24.2	40.3	31.6
1982	439,564	208,074	47.3	0	0.4	2.9	25.6	43.3	27.8
1983	434,570	207,933	47.8	0	0.3	2.7	24.8	43.6	28.6
1984	455,567	216,361	47.5	0	0.2	1.7	20.4	46.5	31.3

Source: Adapted from MAFF. The Meat Statistics of Japan. Tokyo, 1985.

Appendix 6.1. Quantitative description of average Wagyu cow/calf operation, Kagoshima, Japan, 1980 and 1983

Item	Units	1980	1983
Family workers engaged in	Total Persons/farm	2.1	2.0
Wagyu cow/calf operation	Males/farm	1.0	1.0
	Females/farm	1.1	1.0
Area of operation			
Cultivated			
Paddy fields	a/farm[a]	83.9	78.6
Upland fields	"	77.5	79.8
Forage production	"	72.1	65.2
Subtotal	"	161.4	158.4
Livestock			
Buildings, ground	"	3.5	3.1
Forage production	"	--	--
Pasture	"	--	--
Forest and other	"	53.1	49.9
Total	"	218.0	211.4
Other livestock operation			
Hogs		--	--
Dairy cattle		--	--
Broilers		--	--
Hens for eggs		0.7	--
Number of cows	Head/farm	2.5	3.0
Breeding cows			
Average age	Months	71.4	86.9
Live weight	Kg/cow	464.3	466.3
Value	Yen/cow	451,071	501,324
Buildings, facilties and			
implements building area	M^2/farm	67.3	69.6
Cutter	Per farm	1.1	1.1
Hammer mill	"	0.1	0.2
Fodder mixer	"	--	--
Pickup truck	"	--	--
Tractor (implement)	"	--	--
Outputs			
Main products (calves)			
Calves sold	Head/farm	2.1	2.6
Live weight	Kg/calf	261.0	257.5
Value	Yen/calf	391,484	229,577
By-products			
Amount of total manure	Kg/calf	15,117	16,633
Amount of manure utilized	Kg/calf	13,399	16,154
Value of utilized manure	Yen/calf	41,069	55,172

Source: Adapted from MAFF, Production Cost. Tokyo, 1981, pp. 96, 97 and 1984, pp. 94, 95.

[a]One a = $\frac{1}{100}$ hectare or 100 M^2.

Appendix 6.2. Labor use per Wagyu calf produced, average cow/calf operation, Kagoshima, Japan, 1983

Item	1980	1983
	- - - - Hours/calf - - - -	
Preparation and feeding of fodder and water	138.5	107.0
Preparation of bedding and manure removal	59.4	58.6
General care and herding on pasture	19.7	13.8
Other	22.6	13.2
Total	240.2	192.6
Family workers		
Male	151.2	125.5
Female	89.0	67.1
Total	240.2	192.6

Source: Adapted from MAFF, Production Cost. Tokyo, 1981, pp. 96, 97, 98, 99 and 1984, pp. 96, 97.

Appendix 6.3. Production costs for an average Wagyu cow/calf operation, Kagoshima, Japan, 1980 and 1983

Cost item	Cost per calf 1980	Cost per calf 1983	Table 6.1 category
	- - - - Yen	- - - -	
Breeding	19,672	12,622	Cash, nonfeed
Feedcost			
Commercial feed			
Purchased	55,221	56,634	
Self-supplied	18,176	28,019	
Subtotal	73,397	84,653	
Forage crops, pasturage and mowing (self-supplied)	117,545	123,017	
Total	190,942	207,670	Cash, feed
Bedding			
Purchased	2,649	5,847	
Self-supplied	6,905	10,297	
Total	9,554	16,144	Cash, nonfeed
Utilities			
Purchased	1,964	2,320	
Self-supplied	—	—	
Total	1,964	2,320	Cash, nonfeed
Veterinary or medicine	8,828	8,393	Cash, nonfeed
Charges and fees	3,216	5,347	Cash, nonfeed
Depreciation of brood cows	50,699	58,151	Economic, depreciation
Buildings			
Depreciation	7,322	6,938	Economic, depreciation
Repair expenses			
Purchased	2,270	—	
Self-supplied	—		
Subtotal	2,270	—	Cash, nonfeed
Total	9,592	6,938	
Agricultural implements costs			
Depreciation	2,739	2,679	Economic, depreciation
Repairs			
Purchased	160	722	
Self-supplied	—	—	
Subtotal	160	722	Cash, nonfeed
Purchased vs. self-supplied			
Purchased	1,129	1,004	
Self-supplied	—	—	
Subtotal	1,129	1,004	Cash, nonfeed
Total	4,028	4,405	
Labor			
Self-supplied	141,703	132,719	
Hired	—	—	
Total	141,703	132,719	Economic, labor
Total cost			
Purchased	95,109	92,889	
Self-supplied	284,329	294,052	
Depreciation	60,760	67,768	
Total	440,198	454,709	
By-product sales			
Manure	-41,069	-55,172	
Total	-41,069	-55,172	Cash cost, manure credit
Primary production cost (Total cost minus by-products sales)	399,129	399,537	
Capital interest	30,844	34,826	Cash, fixed
Rand rent	15,062	22,795	Cash, fixed
Secondary production cost	445,035	457,158	

Source: Adapted from MAFF, <u>Production Cost</u>. Tokyo, 1981, pp. 100, 101 and 1984, pp. 98, 99.

Appendix 6.4. Cow/calf production costs, per cow, all sizes, Great Plains, 1983

Item	1983
	- Dollars -
Cash receipts:	
Steer calves (1.045 cwt)[1]	71.51
Heifer calves (.820 cwt)	46.87
Yearling steers (1.213 cwt)	70.70
Yearling heifers (.843 cwt)	42.90
Cull cows (.622 cwt)	20.58
Total	252.56
Cash expenses	
Feed	
Grain (1.816 cwt)	9.27
Silage (.373 ton)	8.05
Protein supplements (2.290 cwt)	27.10
Salt and minerals (.349 cwt)	1.68
Hay (1.076 tons)	32.34
Pasture (4.171 AM)	29.43
Private range (8.028 AUM)	0
Public grazing (.342 AM)	.74
Crop residue (.474 AUM)	.21
Other	
Veterinary and medicine	6.04
Livestock hauling	1.57
Marketing	4.68
Custom feed mixing	.53
Fuel, lube, and electricity	20.45
Machinery and building repairs	19.78
Hired labor (3.063 hours)	11.85
Total, variable expenses	173.72
General farm overhead	14.19
Taxes and insurance	18.15
Interest	44.76
Total, fixed expenses	77.10
Total, cash expenses	250.82
Economic costs	
Variable expenses	173.72
General farm overhead	14.19
Taxes and insurance	18.15
Capital replacement	59.61
Allocated returns to owned inputs	
Operating capital	7.66
Other nonland capital	43.75
Land	128.39
Unpaid labor (17.132 hours)	67.10
Total, economic costs	512.57
Residual to management and risk	-260.01
Net returns to owned inputs	-13.11

Source: USDA. Economic Indicators of the Farm Sector: Cost of Production, 1983. ERS, ECIFS 3-1, Washington, July, 1984, p. 111.

Appendix 6.5. Quantitative description of 16 head Wagyu fattening operation, Kyushu, Japan, 1980 and 1983

Item	Units	1980	1983
Family workers engaged in	Total/farm	2.6	2.4
Wagyu fattening	Males/farm	1.4	1.2
	Females/farm	1.2	1.2
Area of operation			
Cultivated			
Paddy fields	a/farm	84.0	82.2
Upland fields	"	82.6	81.8
Subtotal	"	166.6	164.0
Livestock			
Buildings, ground	"	4.2	4.6
Forage production	"	3.1	0.5
Pasture	"	0.0	6.8
Forest and other	"	88.6	92.6
Total	"	262.5	268.5
Other livestock operations			
Hogs	Head/farm	0.4	0.0
Dairy cattle		0.6	0.5
Broilers		—	—
Hens for eggs		0.2	0.4
Number of cattle raised per month	Head/farm	13.2	16.3
Original cattle			
Average months	Months old	10.4	9.8
Live weight	Kg/head	297.6	293.7
Value	Yen/head	299,618	342,231
Fattening period	Months/head	17.2	17.4
Weight increased by fattening	Kg/head	317.2	335.8
Buildings, facilities and implements			
Area of buildings	M^2/farm	161.3	186.1
Automatic water feeder	Per farm	0.1	0.4
Cutter	"	0.9	0.8
Hammermill	"	0.2	0.0
Fodder mixer	"	0.0	0.0
Pickup truck	"	0.5	0.7
Tractor (implement)	"	0.4	0.3
Outputs			
Main products			
Finished cattle sold	Head/farm	9.2	11.6
Live weight	Kg/head	606.9	629.2
Value	Yen/head	620,061	655,176
By-products			
Amount of total manure	Kg/head	11,611	12,366
Amount of manure utilized	Kg/head	10,036	10,918
Value of utilized manure	Yen/head	19,873	26,600

Source: Adapted from MAFF, Production Cost. Tokyo, 1981, pp. 108, 109 and 1984, pp. 106, 107.

Appendix 6.6. Labor hours per Wagyu slaughter steer fed, average feeding operation, Kyushu, Japan, 1980 and 1983

Item	1980	1983
	- - - Hours/head - - -	
Preparation and feeding of fodder and water	58.0	53.3
Preparation of bedding and manure removal	22.7	18.8
General care, pasturing	4.1	3.3
Other	6.1	4.3
Total	90.9	79.7
Family workers		
Male	62.4	49.1
Female	28.5	29.7
Subtotal	90.9	78.8
Hired workers		
Subtotal	0.0	0.9

Source: Adapted from MAFF, Production Cost. Tokyo, 1981, pp. 110, 111 and 1984, pp. 108, 109.

Appendix 6.7. Production costs on an average Wagyu fattening operation, Kyushu, Japan, 1980 and 1983

Cost item	Cost per head 1980	Cost per head 1983	Table 6.1 category
	- - - - Yen	- - - -	
Original livestock	302,227	346,390	Cash, feeder cattle
Feedcost			
Commercial feed			
Purchased	188,014	209,531	
Self-supplied	5,637	6,390	
Subtotal	193,651	215,921	
Forage crops, pasturage and mowing (self-supplied)	27,980	25,595	
Total	221,631	241,516	Cash, feed
Bedding			
Purchased	2,049	10,761	
Self-supplied	8,018	1,692	
Total	10,067	12,453	Cash, nonfeed
Utilities			
Purchased	1,951	3,491	
Self-supplied	--	19	
Total	1,951	3,510	Cash, nonfeed
Veterinary or medicine	2,613	2,952	Cash, nonfeed
Charges and fees	584	523	Cash, nonfeed
Buildings			
Depreciation	7,472	7,386	Ownership, depreciation
Repair expenses			
Purchased	487	270	
Self-supplied	89	20	
Subtotal	576	290	
Total	8,048	7,676	Cash, nonfeed
Agricultural implements costs			
Depreciation	5,143	6,650	Ownership, depreciation
Repairs			
Purchased	912	1,010	
Self-supplied	51	18	
Subtotal	963	1,028	Cash, nonfeed
Purchased vs. self-supplied			
Purchased	455	312	
Self-supplied	4	4	
Subtotal	459	316	Cash, nonfeed
Total	6,565	7,994	
Labor			
Self-supplied	62,293	60,173	
Hired	7	763	
Total	62,300	60,936	Economic cash, nonfeed
Total cost			
Purchased	499,299	575,552	
Self-supplied	104,072	94,362	
Depreciation	12,615	14,036	
Total	615,986	638,950	
By-product sales			
Manure	-19,873	-26,600	
Total	-20,525	-27,908	Manure sales
Primary production (Total cost minus by-products sales)	595,461	656,042	
Capital interest	18,710	18,735	Cash, interest
Rand rent	4,345	5,627	Cash, land
Secondary production cost	618,516	680,404	

Source: Adapted from MAFF, Production Cost. Tokyo, 1981, pp. 112, 113 and 1984, pp. 110, 111.

Appendix 6.8. Fed cattle production costs in the United States, 1983

Item	Farmer feedlots	Commercial feedlots	All size feedlots
	---------- $Cwt ----------		
Cash receipts			
Fed beef (100 lbs)	62.03	62.15	62.12
Cash expenses			
Feeder cattle	35.21	38.30	37.62
Feed			
Haylage	0.31	0.00	0.07
Silage	2.68	0.47	0.96
Dry grain	9.00	13.82	12.75
Concentrates	2.64	0.00	0.58
Legume hay	0.54	1.60	1.37
Pasture	0.01	0.00	0.00
Other roughages	0.00	0.44	0.34
Crop residue	0.00	0.00	0.00
Protein supplement	3.00	4.59	4.24
Other			
Veterinary and medicine	0.54	1.10	0.98
Livestock hauling	0.70	0.45	0.51
Marketing	0.32	0.18	0.21
Bedding	0.22	0.00	0.05
Legal and accounting fees	0.00	0.26	0.20
Fuel, lube and electricity	1.25	1.53	1.47
Machinery and building repairs	1.55	0.30	0.58
Hired labor	0.52	0.68	0.64
Miscellaneous	0.00	0.35	0.27
Manure credit	-0.11	-0.10	-0.10
Total, variable expenses	58.38	63.97	62.74
General farm overhead	1.19	0.00	0.30
Taxes and insurance	0.52	0.08	0.18
Hired management	0.00	0.18	0.14
Interest	4.64	4.08	4.22
Total, fixed expenses	6.35	4.34	4.84
Total, cash expenses	64.73	68.31	67.58
Economic costs			
Variable expenses	58.38	63.97	62.74
General farm overhead	1.19	0.00	0.30
Taxes and insurance	0.52	0.08	0.18
Capital replacement	3.62	0.52	1.21
Hired management	0.00	0.18	0.14
Allocated returns to owned inputs			
Operating capital	1.71	1.20	1.31
Other nonland capital	1.43	0.21	0.48
Land	0.66	0.06	0.20
Unpaid labor	2.37	0.00	0.52
Total, economic costs	69.88	66.22	67.08

Source: USDA. Economic Indicators of the Farm Sector: Cost of Production, 1983. ERS, ECIFS 3-1, Washington, July, 1984, pp. 105, 106, 107.

Appendix 6.9. Data source to Tables 6.5 and 6.6 on estimated costs and returns for a 50 head dairy steer finishing operation in Hokkaido, Japan, 1980 and 1983

Item	1980	1983
	- - - - - - - - ¥ - - - - - - - -	
Investment		
Land[a]	5,360,000	7,622,800
Buildings and capital investment[b]	14,061,800	13,159,894
Equipment[c]	1,403,500	1,503,290
Cattle[d]	--	--
Total	20,825,300	22,285,984
Basic production costs		
Cattle[d]	9,466,800	8,943,350
Feedstuffs, nonpasture[e]	9,502,594	9,851,575
Feedstuffs, pasture[f]	218,438	342,945
Bedding[g]	357,984	513,354
Veterinary service and medicine		139,937
Maintenance[i]	156,797	101,575
Utilities[j]	156,797	153,281
Other[k]	26,953	38,438
Total less cattle	10,558,407	11,141,105
Total with cattle	20,025,207	20,084,455
By-product value[l]	379,313	424,913
Net basic production cost		
Total less cattle	10,179,094	10,716,192
Total with cattle	19,645,894	19,659,542
Other costs		
Capital costs		
Interest on operating capital (10% of 60% (of production with cattle cost)[m]	1,201,512	1,205,067
Interest in buildings and capital investment[n]	843,708	789,594
Interest on equipment[n]	--	90,197
Subtotal	2,129,430	2,084,858

Continued

Appendix 6.9. Data source to Tables 6.5 and 6.6 on estimated costs and returns for a 50 head dairy steer finishing operation in Hokkaido, Japan, 1980 and 1983 (continued)

Item	1980	1983
	- - - - - - - - ¥ - - - - - - - -	
Ownership costs		
Depreciation[o]	1,225,973	1,179,782
Taxes[p]	208,253	--
Insurance[q]	--	--
Subtotal	1,434,226	1,179,782
Land rent	--	--
Labor[r]	1,155,516	1,266,000
Management[s]	--	--
Subtotal	1,155,516	1,266,000
Total, other costs	4,719,172	4,530,640
Total net basic production and other costs (less cattle)	14,898,266	15,716,192
Total net basic production and other costs (with cattle)	24,365,066	24,190,182
Income[t]	24,402,495	23,278,345
Net income		
Above basic production costs	4,756,601	3,618,803
Above all costs	37,429	-911,837
Breakeven (calculations for 1983)[u]		
Net basic production costs (¥19,659,542 ÷ 31,463 kg)	654	625
All costs (¥15,716,192 ÷ 17,687 kg)	811	769
Cost of gain (calculation for 1983)[v]		
Net basic production cost (¥10,716,192 ÷ 17,687 kg)	576	606
All costs (¥15,716,192 ÷ 17,687 kg)	843	888

Source: Appendix 6.10 for 1983 expenses. Data for 1980 are not provided.

Appendix 6.10. Footnotes to Appendix 6.9 on estimated costs and returns for a 50 dairy steer finishing operation in Hokkaido, Japan, 1983

[a]Land

Item	a	m^2	Total m^2
			- ¥ -
Forage production	1.0	0.08	80
Buildings and surrounding area	31.5	0.315	3,150
			3,230

at ¥2,360/m^2 = ¥7,622,800 total cost

[b]Buildings and capital investment

Item size		Cost per m^2	Total cost
	- m^2 -	- - - - - ¥ - - - - -	
Buildings	250	32,620	8,155,000
Manure storage	180	12,582	2,264,760
Bedding storage and forage drying	50	27,960	1,398,000
Working facilities	78	11,277	879,606
Scale (1.5 ton)	--	--	462,528
			13,159,894

[c]Equipment

Item	Total value new	Allocation to livestock	Total cost
	- - ¥ - -	- - % - -	- - ¥ - -
Tractor (67ps)	1,080,890	100	1,080,890
Cutter (10ps)	422,400	100	422,400
			1,503,290

Continued

Appendix 6.10. Footnotes to Appendix 6.9 on estimated costs and returns for a 50 dairy steer finishing operation in Hokkaido, Japan, 1983 (continued)

[d]Cattle

Shown as a production cost. 50 head weighing 271.6 kg at ¥178,867 each at 7.6 months of age to be fattened for 12.7 months. Total weight is 271.6 x 50 = 13,580 kg. The cost is ¥178,867 x 50 = ¥8,943,350.

[e]Feedstuffs, nonpasture

Purchased	¥207,860
Farm grown	865
Total per steer	208,525
Total for 50 steers	¥10,426,250

This is for 12.7 months, so the cost is ¥9,851,575 per year.

Calculations of purchased feedstuffs in footnote f

Item	Purchased, per dairy steer		Self-grown, per dairy steer	
	Amount	Value	Amount	Value
	- kg -	- ¥ -	- kg -	- ¥ -
Grains				
Cracked rice	32	1,783	0.0	--
Barley	428	24,293	0.0	--
Wheat	309	15,711	0.0	--
Other	--	439	--	--
Subtotal	--	42,226	--	--
Bran				
Wheat bran	175	7,891	--	--
Rice bran	14	475	0.0	1
Subtotal	--	8,652	--	1
Nonannual dregs				
Soybean dregs	7	666	--	--
Tofu dregs	2	32	--	--
Other	--	4,757	--	--
Subtotal	--	5,455	--	--

Continued

Appendix 6.10. Footnotes to Appendix 6.9 on estimated costs and returns for a 50 dairy steer finishing operation in Hokkaido, Japan, 1983 (continued)

Calculations of purchased feedstuffs in footnote f (continued)

Item	Purchased, per dairy steer		Self-grown, per dairy steer	
	Amount	Value	Amount	Value
	- kg -	- ¥ -	- kg -	- ¥ -
Combination fodder				
For adult combination fodder	2,157	127,631	--	--
For young combination fodder	188	11,458	--	--
Subtotal	2,345	139,089	--	--
Skim-milk				
Skim-milk	0	19	--	--
Artificial milk	1	225	--	--
Other	--	585	--	--
Subtotal	1	829	--	--
Potatoes and vegetables	--	45	--	--
Straw				
Rice straw	231.0	7,889	27	643
Other	--	1,720	--	21
Subtotal	231.0	9,609	--	664
Other				
Salt	3	514	--	--
Calcium	3	180	--	--
Other	--	1,261	--	--
Total	--	207,867	--	665

[f] Feedstuffs, pasture

¥7,259 per steer x 50 steers = ¥362,950 for 12.7 months.

¥342,945 for 1 year.

Continued

Appendix 6.10. Footnotes to Appendix 6.9 on estimated costs and returns for a 50 dairy steer finishing operation in Hokkaido, Japan, 1983 (continued)

[g]Bedding

Item	Amount	Cost per unit	Total cost
	- kg -	- - - - - ¥ - - - - -	
Rice straw	2.0	16.7	334
Sawdust	1,417	5.2	7,426
Other	—		3,106
Total per steer			10,866

For 50 steers = ¥543,000 for 12.7 months and ¥513,354 for 1 year.

[h]Veterinary service and medicine

¥2,962 per steer x 50 steers = ¥148,100 for 12.7 months or ¥139,937 for 1 year.

[i]Maintenance

This includes supplemental purchase as well as repairs but not depreciation.

Machinery	¥1,682
Buildings	¥468
Total per steer	¥2,150

For 50 steers = ¥107,500 for 12.7 months and ¥101,575 for 1 year.

[j]Utilities

Electricity, water, fuel (incl. for vehicle) and materials used in power production. ¥3,270 per steer x 50 steers = ¥163,500 for 12.7 months or ¥153,281 for 1 year.

Continued

Appendix 6.10. Footnotes to Appendix 6.9 on estimated costs and returns for a 50 dairy steer finishing operation in Hokkaido, Japan, 1983 (continued)

---

[k]Other

Building and machinery rent, manure transport, animals registration fee, disease testing fee, and material and labor cost related to these items. ¥820 per steer x 50 steers = ¥41,000 for 12.7 months or ¥38,438 for 1 year.

---

[l]By-product value

The amount of manure produced is 8,385 kg of which 5,725 kg is utilized per steer and its value is ¥8,994.
¥8994 x 50 = ¥449,700 for 12.7 months or ¥424,913 for 1 year.

---

[m]Operating capital

10 percent interest. Operating capital calculated on 60 percent of the total with cattle production cost because about half of total cost is in cattle which are purchased at the beginning. Other costs take place continuously during the year.

---

[n]Interest on buildings, capital investment and equipment

6 percent annual interest.

Continued

Appendix 6.10. Footnotes to Appendix 6.9 on estimated costs and returns for a 50 dairy steer finishing operation in Hokkaido, Japan, 1983 (continued)

[o]Depreciation

Item	Total cost	Depreciation factor	Depreciation
	- - ¥ - -	- % -	- - ¥ - -
Buildings and capital investment			
Buildings	8,155,000	6	489,300
Manure storage	2,264,760	10	226,476
Bedding storage	1,398,000	7	97,860
Working facilities	879,606	10	87,961
Scale	462,528	18	83,255
Subtotal			984,852
Equipment			
Tractor	1,080,890	11	118,898
Cutter	422,400	18	76,032
Subtotal			194,930
Total			1,179,782

[p]Taxes

None because considered as a payment from net income in the Japanese accounting system.

[q]Insurance

Covered by veterinary expense and maintenance fee.

[r]Labor

¥26,797 per steer fed x 50 steers = ¥1,339,850 for 12.7 months or ¥1,266,000 for 1 year.

Continued

Appendix 6.10. Footnotes to Appendix 6.9 on estimated costs and returns for a 50 dairy steer finishing operation in Hokkaido, Japan, 1983 (continued)

---

[r]Labor (continued)

The total hours are 27.9 per steer fed for 12.7 months or 26.4 hours for 1 year. Of the 27.9 hours, 17.7 are spent in preparing feedstuffs, feeding and watering, 6.3 hours in adding bedding and manure removal, and 0.7 hours in general care. 26.9 hours of the 27.9 hours are from the family. Out of 27.9 hours, 18.4 hours of the family labor are by males.

---

[s]Management

Included in labor.

---

[t]Income

Death loss of one percent in 12.7 months, so an average of 49.5 steers are sold at the end of 12.7 months weighing 672.7 kg at ¥497,702 = ¥739.85 per kg. The total income on an annual basis is ¥497,902 x 49.5 x 12/12.8 = ¥23,278,345.

---

[u]Breakeven

Kilos sold is 33,298.65 (672.7 x 49.5) for 12.7 months or 31,463.20 on an annual basis.

---

[v]Cost of gain

Net kilos sold is 18,719 kg (32,299 - 13,580) for 12.7 months or 17,687 kg for 1 year.

Continued

Appendix 6.10. Footnotes to Appendix 6.9 on estimated costs and returns for a 50 dairy steer finishing operation in Hokkaido, Japan, 1983 (continued)

Sources: by footnote.

[a]Institute of Japan for land cost and MAFF, 1984, pp. 120-121.

[b]Updated from Hokusen, 1982, p. 26.

[c]Updated from Hokusen, 1982, p. 26.

[d]MAFF, 1984.

[e]MAFF, 1984, p. 124.

[f]MAFF, 1984, p. 124.

[g]MAFF, 1984, pp. 130, 131.

[h]MAFF, 1984, p. 125.

[i]MAFF, 1984, pp. 124, 125.

[j]MAFF, 1984, pp. 124, 125.

[k]MAFF, 1984, p. 125.

[l]MAFF, 1984, p. 121.

[m]Own calculations.

[n]Own calculations.

[o]Footnote b and c.

[p]None.

[q]None.

[r]Own calculations.

[s]None.

[t]None.

[u]None.

[v]None.

Appendix 6.11. Data source to table 6.5 on estimated costs for 300 head dairy steer finishing operation in Upper Midwest, USA, 1980 and 1983

Land

0.10 ha at $3,750 per ha = $375

Buildings and capital investment[a]

	1980	1983
Buildings	$15,300	$17,595
Pit walls	12,000	13,800
Pit floor	7,200	8,280
Floor		
Slats	13,660	15,709
Beams	4,200	4,830
Columns	1,200	1,380
Gates	2,100	2,415
Bunks & fence	2,700	3,105
Excavating	3,000	3,450
Loading chute	1,300	1,495
Corrals	4,600	5,290
Waterers	800	920
Plumbing	600	690
Electrical	300	345
Total	68,960	79,304

Equipment[a]

	1980	1983
Waste handling		
Liquid manure tank	$5,000	$5,750
Agitator	4,065	4,675
Tractor (20% usage)	3,000	3,450
Feed handling		
270 bu + 5 ton auger mixer wagon	11,000	12,650
Tractor 40% usage	6,000	6,900
Feed storage		
Facilities & related equip.	32,664	37,564
Total	61,729	70,989

Continued

Appendix 6.11. Data source to table 6.5 on estimated costs for 300 head dairy steer finishing operation in Upper Midwest, USA, 1980 and 1983 (continued)

<u>Cattle</u>

300 head of calves weighing 48.18 kg at $1.98 per kg = $95.40 per head, for 1983, and 2.77 per kg = $95.40 in 1980.

<u>Feedstuffs, nonpasture</u>

2,333 kg per head at $10.1591 per kg = $371.18 per head in 1983, and 0.1446 per kg and $371.18 in 1980.

<u>Veterinary Services and Medicines</u>

$20.28 per head in 1983 and $17.18 per head in 1980.

<u>Maintenance</u>

Adapted from Simpson, Baldwin and Baker, p. 36 and updated.

<u>Utilities</u>

Adapted from Simpson, Baldwin and Baker, p. 36 and updated.

<u>Other</u>

Includes telephone, office supplies, gas, oil, travel and entertainment, bookkeeping, and misc.
Adapted from Simpson, Baldwin and Baker, p. 36 and updated.

<u>By-product value</u>

Value of manure after spreading on fields, $10 per head.

Continued

Appendix 6.11. Data source to table 6.5 on estimated costs for 300 head dairy steer finishing operation in Upper Midwest, USA, 1980 and 1983 (continued)

Interest on operating capital

12 percent on 90 percent of with cattle production costs in 1983 and 16 percent interest in 1980.

Interest on buildings and capital Investment

12 percent in 1983 and 16 percent in 1980.

Interest on Equipment

12 percent in 1983 and 16 percent in 1980.

Depreciation

Five percent on buildings and capital investment.
Ten percent on equipment.

Labor

Includes labor used in feeding, manure disposal and overhead activities. Overhead activities include such items as observation, sorting, handling, buying and selling cattle and feed. About 4.50 hours are required per head or a total of 1,350 hours at $5.50 per hour in 1983 and $5.00 per hour in 1980.

Management

Included in labor cost assuming that all activities are carried out by the owner.

Continued

Appendix 6.11. Data source to table 6.5 on estimated costs for 300 head dairy steer finishing operation in Upper Midwest, USA, 1980 and 1983 (continued)

Income

Cattle gain an average of 1.19 kg per day for the 361 days they are on feed or a total of 429.59 kg per head which, added to the purchase weight of 48.18 kg yields a final sale weight of 477.77 kg. Net selling price is $1.33 per kg or a total of $635.43 per head. Assuming a 1 percent death loss, 297 head are sold for a total income of $188,723. Adapted from Blair, 1983; Burroughs, et.al., 1965; Buroughs, et.al., 1966; and Williams, Topel, Gray and Jetter, 1969.

Breakeven

Costs with cattle divided by kg sold (132,342 kg).

Cost of gain

Cost without cattle divided by net gain (118,996 kg).

---

[a]1983 is updated from McGrann, et al, 1979.

Appendix 8.1. Potential national level productivity improvement and kilos of beef produced in beef breed weaned calves by adoption of various techniques and technologies in Japan, 1990, 2000 and 2010

Item	Description of benefits	Potential productivity improvement by			Additional kilos produced per 100 cows[a]		
		1990	2000	2010	1990	2000	2010
		- - - Percent - - -			- - - - - - - Kilos - - - - - -		
		Breeding					
Sire evaluation programs	1) Increased weaned calf weight	0-1	1-2	2-4	0-221	221-442	442-867
Cow selection by reproduction records	1) Increased weaned calf weight	0-0	0-1	0-1	0-0	0-221	0-221
	2) Increased weaned calf crop	0-0	1-2	2-3	0-0	221-442	442-663
	3) Improved cow milking performance	0-1	0-1	1-2	0-221	0-221	221-442
Crossbreeding for hybrid vigor	1) Increased weaned calf weight	0-0	0-1	1-2	0-0	0-221	221-442
Subtotal		0-2	2-7	6-12	0-442	442-1,768	1,326-2,635
		Reproduction					
Reduced calving intervals by improved heat detection and pregnancy testing	1) Increased calving percent	1-2	2-3	3-4	210-420	420-630	630-824
Embryo transfer for genentic improvement	1) Increased weaned calf rate	0-0	0-1	2-3	0-0	0-221	442-663
	2) Improved cow milking performance	0-0	0-1	0-1	0-0	0-221	0-221
	3) Improved calving percent	0-0	0-1	0-1	0-0	0-210	0-210
Embryo transfer for multiple births	1) Increase calving percent	0-0	1-2	2-3	0-0	210-420	410-630
Sexing embryos	1) Greater proporton of male calves	0-0	0-1	0-2	0-0	0-221	0-442
Genetic engineering for rapid growth	1) Increased weaned calf weight	0-0	0-1	0-2	0-0	0-221	0-442
Subtotal		1-2	3-10	7-16	210-420	840-2,144	1,482-3,432

Continued

Appendix 8.1. Potential national level productivity improvement and kilos of beef produced in beef breed weaned calves by adoption of various techniques and technologies in Japan, 1990, 2000 and 2010 (continued)

Item	Description of benefits	Potential productivity improvement by 1990	2000	2010	Additional kilos produced per 100 cows[a] 1990	2000	2010
		- - - Percent - - -			- - - - - - - Kilos - - - - - -		
			Nutrition				
Optimal ration to calf	1) Increased weaning weight	0–0	0–1	0–2	0–0	0–221	0–442
Optimal ration to cow	1) Increased calving percent	0–0	0–1	1–2	0–0	0–210	210–420
Treatment of straw and hay by alkali, exploding or other methods	1) Increased weaning weight	1–2	2–3	3–4	221–442	442–663	663–867
Public land for grazing	1) Reduced feed costs[c]	0–0	1–2	1–3	0–0	0–0	0–0
Access to less expensive nationally produced feed	1) Reduced feed costs and increased weaning weight	0–0	0–1	1–2	0–0	0–221	221–442
Subtotal		1–2	3–8	6–13	221–442	442–1,315	1,094–2,171
			Management				
Death loss	1) Increased weaned calf crop	0–1	0–1	0–1	0–221	0–221	0–221
Growth stimulating implants (Ralgro)	1) Increased weaning weight	0–2	2–3	2–3	0–442	442–663	442–663
Large size drylots (75–150 cows)	1) Reduced costs and increased calf crop and weight	0–1	2–3	2–3	0–221	442–663	442–663
Record keeping and analysis	1) Increased weaned calf weight and calf crop	2–3	3–4	3–4	442–663	663–867	663–867
Improved access to new knowledge	1) Increased weaned calf weight and calf crop	0–1	1–2	1–2	0–221	221–442	221–442
Improved culling practices	1) Increased weaned calf weight and calf crop	0–1	1–2	3–4	0–221	221–442	663–867
General management and other	1) Increased weaned calf weight and calf crop	0–1	1–2	2–3	0–221	221–442	442–663
Subtotal		2–10	10–17	13–20	442–2,210	2,210–3,740	2,873–4,386
Total					873–3,514	3,934–8,967	6,775–12,624
Mean					2,194	6,451	9,700
					- - - - - - Percent - - - - - -		
Percent increase in kilos of weaned calf[b]							
Total					04–15	17–39	30–55
Mean					10	28	43

[a]See Appendix 8.5 for conversion factors from productivity increases as percentages to kilos produced per 100 cows.

[b]Derived using a weaning weight of 285 kg times 80 cows (percent weaning a calf) equals 22,800 kg as a base for 1984.

[c]No impact on productivity.

Appendix 8.2. Potential national level productivity improvement and kilos of dairy calves from adoption of various techniques and technologies in Japan, 1990, 2000 and 2010

Item	Description of benefits	Potential productivity improvement by			Additional kilos produced per 100 cows[a]		
		1990	2000	2010	1990	2000	2010
		- - - Percent - - -			- - - - - - - - Kilos - - - - - -		
			Breeding				
Sire evaluation programs	1) Increased weaned calf weight	0–0	0–0	0–0	0–0	0–0	0–0
Cow selection by reproduction records	1) Increased weaned calf weight	0–0	0–0	0–0	0–0	0–0	0–0
	2) Increased weaned calf crop	0–0	0–0	0–0	0–0	0–0	0–0
Subtotal		0–0	0–0	0–0	0–0	0–0	0–0
			Reproduction				
Reduced calving intervals by improved heat detection and pregnancy testing	1) Increased calving percent	0–1	1–2	1–2	0–210	210–420	210–420
Embryo transfer for multiple births	1) Increased calving percent	0–1	3–8	4–10	0–210	630–1,648	824–2,116
Embryo transfer of beef breeds	1) Increased final calf weight	0–1	2–3	3–4	0–221	442–663	663–867
Sexing embryos	1) Produce greater proportion of male calves	0–0	0–1	0–1	0–0	0–221	0–221
Genetic engineering for rapid growth	1) Increased final calf weight	0–0	0–1	1–2	0–0	0–221	221–442
Subtotal		0–3	6–15	9–19	0–641	1,282–3,173	1,918–4,066

Continued

Appendix 8.2. Potential national level productivity improvement and kilos of dairy calves from adoption of various techniques and technologies in Japan, 1990, 2000 and 2010 (continued)

Item	Description of benefits	Potential productivity improvement by 1990	2000	2010	Additional kilos produced per 100 cows[a] 1990	2000	2010
		- - - Percent - - -			- - - - - - - Kilos - - - - - -		
			Nutrition				
Optimal calf ration	1) Regulation of concentrate in ration (increased final calf weight)	0–1	1–2	1–2	0–221	221–442	221–442
Access to less expensive nationally produced feed	1) Reduced feed costs and increased weaning weight	0–0	0–1	1–2	0–0	0–221	221–442
Subtotal		0–1	1–3	2–4	0–221	221–663	442–884
			Management				
Death loss	1) Increased percentage of raised calves	0–1	1–2	1–2	0–221	221–442	221–442
Cooperative nursery	1) Reduced death loss (see above)	—	—	—	—	—	—
	2) Faster gains (increased weight)	0–1	1–2	2–3	0–221	221–442	442–663
	3) Computerized culling of probable poor gaining calves in the feedlot phase (increased calf weight)	0–0	0–1	1–2	0–0	0–221	221–442
Integrated nursery, calf rearing, feedlot	1) Faster gains (increased weight)	0–0	0–1	1–2	0–0	0–221	221–442
Growth stimulating implants (Ralgro)	1) Increased final calf weight	0–2	2–3	2–3	0–442	442–663	442–663
Record keeping and analysis	1) Increased final calf weight and percentage of calves	1–2	2–3	2–3	221–442	442–663	442–663
Improved access to new knowledge	1) Increased final calf weight and percentage of calves	0–0	0–1	0–1	0–0	0–221	0–221
General management and other	1) Increased final calf weight and percentage of calves	0–1	1–2	1–2	0–221	221–442	221–442
Subtotal		1–7	7–15	10–18	221–1,547	1,547–3,315	2,210–3,978
Total					221–2,049	3,050–7,151	4,570–8,928
Mean					1,135	5,101	6,749
					- - - - - - Percent - - - - - -		
Percent increase in kilos of calf available to feedlots							
Total					01–09	13–31	20–39
Mean					05	22	30

[a]See Appendix 8.5 for conversion factors from productivity increases as percentages to kilos produced per 100 cows.

[b]Derived using a weaning weight of 272 kg of calf entering the feedlot times 85 cows (percent weaning a calf) equals 23,120 kg as a base for 1984.

Appendix 8.3. Potential national level productivity improvement in beef breed feedlots from adoption of various techniques and technologies in Japan, 1990, 2000 and 2010

Item	Description of benefits	Potential productivity improvement by 1990	2000	2010
		- - - - -	Percent	- - - - -
	Breeding			
Sire evaluation programs	1) Increased growth rate[a]	0–1	1–2	1–2
	2) Increased marbling score[c]	---	---	---
	3) Increased feed efficiency[b]	0–1	1–2	2–3
Cow selection by reproduction records	1) Increased growth rate[a]	0–0	0–1	0–2
	2) Increased marbling score[c]	---	---	---
	3) Improved feed efficiency[b]	0–0	0–1	0–2
Crossbreeding for hybrid vigor	1) Increased growth rate[a]	0–0	1–2	3–5
	2) Increased marbling score[c]	---	---	---
	3) Increased feed efficiency[b]	0–0	0–1	1–2
	Reproduction			
Sexing embyros	1) Increased proportion of males so increased growth rate[a]	0–0	0–1	1–2
Genetic engineering for rapid growth	1) Increased growth rate[a]	0–0	0–1	1–2
	Nutrition			
Rumen regulator	1) Improved feed conversion[b]	1–2	2–3	2–3
Abortants to heifers	1) Improved efficiency[b]	0–0	0–1	0–1
Growth stimulating implants (Ralgro)	1) Increased daily gain[a] (8–10 percent potential)	0–2	2–3	2–3
Improved silage and storage methods	1) Reduced feed costs[c]	1–2	2–3	3–4
Efficient use of by-products and animal wastes	1) Reduced feed costs[c]	1–2	2–3	3–4
Access to less expensive nationally produced feed	1) Reduced costs of feedstuffs[c] (in addition to other items evaluated)	0–1	1–2	2–3

Continued

Appendix 8.3. Potential national level productivity improvement in beef breed feedlots from adoption of various techniques and technologies in Japan, 1990, 2000 and 2010 (continued)

		1990	2000	2010
	Management			
Decision to shorten the fattening period	1) Increased growth rate[a]	2–3	3–4	3–4
	2) Reduced total quantity of inputs, and gain in particular	—	—	—
Feed to lighter weights	1) Reduced cost of gain[c]	1–2	2–3	2–3
Large scale feedlots, perhaps cooperative	1) Economies of size[c]	0–1	1–2	1–2
	2) Better management (see other items)	—	—	—
Reduced death loss	1) Reduced cost of gain[c]	0–1	0–1	0–1
Grade estimation of live animal by ultrasonic estimation	1) Reduced cost of gain by slaughter at optimal age[c]	0–0	0–0	0–1
Waste management	1) Improved daily gain[a]	0–0	0–1	0–1
Record keeping and analysis (inc. computer use)	1) Improved daily gain[a]	0–1	1–2	1–2
Improved access to new knowledge	1) Improved daily gain[a]	0–0	0–1	0–1
Improved efficiency in animal and feed procurement and transportation	1) Reduced feed and animal cost[c]	0–0	0–1	0–1
General management and other	1) Improved average daily gain[a]	0–1	0–1	0–1
Growth rate (sum of footnote a)				
Total		2–8	8–18	12–24
Mean		5	13	18
Feed efficiency (sum of footnote b)				
Total		1–3	3–8	5–11
Mean		2	6	8
Other economic efficiency (sum of footnote c)				
Total		3–9	8–16	11–20
Mean		6	12	16

Appendix 8.4. Potential national level productivity improvement in dairy breed feedlots from adoption of various techniques and technologies in Japan, 1990, 2000 and 2010

Item	Description of benefits	Potential productivity improvement by 1990	2000	2010
		-----	Percent	-----
	Breeding			
Sire evaluation programs	1) Increased growth rate[a]	0–0	0–0	0–0
	2) Increased marbling score	0–0	0–0	0–0
	3) Increased feed efficiency[b]	0–0	0–0	0–0
	Reproduction			
Sexing embyros	1) Increased proportion of males[a]	0–0	0–1	1–2
Genetic engineering for rapid growth	1) Increased growth rate[a]	0–0	0–1	1–2
	Nutrition			
Rumen regulator	1) Improved feed conversion[b]	1–2	2–3	2–3
Abortants to heifers	1) Improved efficiency[b]	0–0	0–0	0–0
Growth stimulating implants (Ralgro)	1) Increased daily gain[a] (8-10 percent potential)	0–2	2–3	2–3
Improved silage and storage methods	1) Reduced feed costs[c]	1–2	2–3	3–4
Efficient use of by-products and animal wastes	1) Reduced feed costs[c]	1–2	2–3	3–4
Access to less expensive nationally produced feed	1) Reduced costs of feedstuffs[c] (in addition to other items evaluated)	0–1	1–2	2–3

Continued

Appendix 8.4. Potential national level productivity improvement in dairy breed feedlots from adoption of various techniques and technologies in Japan, 1990, 2000 and 2010 (continued)

	Management			
Decision to shorten the fattening period	1) Reduced cost of gain[c]	1-2	2-3	2-3
	2) Reduced total quantity of inputs, and grain in particular[b]	---	---	---
Feed to lighter weights	1) Reduced cost of gain[c]	1-2	2-3	2-3
Large scale feedlots, perhaps cooperative	1) Economies of size[c]	0-1	1-2	1-2
	2) Better management (see other items)	---	---	---
Reduced death loss	1) Reduced cost of gain[c]	0-1	0-1	0-1
Grade estimation of live animal by ultrasonic estimation	1) Reduced cost of grain by slaughter at optimal age[c]	0-0	0-0	0-0
Waste management	1) Improved daily gain[a]	0-0	0-1	0-1
Record keeping analysis by computer	1) Improved daily gain[a]	0-1	1-2	1-2
Improved access to new knowledge	1) Improved daily gain[c]	0-0	0-1	0-1
Improved efficiency in animal and feed procurement and transportation	1) Reduced feed and animal cost[c]	0-0	0-1	0-1
General management and other	1) Improved average daily gain[a]	0-1	0-1	0-1
Growth rate (sum of footnote a)				
Total		0-4	3-9	5-11
Mean		2	6	8
Feed efficiency (sum of footnote b)				
Total		1-2	2-3	2-3
Mean		1	2	2
Other economic efficiency (sum of footnote c)				
Total		4-11	10-19	13-22
Mean		8	15	18

Appendix 8.5. Conversion fctors from productivity improvement percentages to additional kilos of live animal produced for beef breed cow/calf operations and dairy calves[a]

Increased weaned calf weight[b]		Increased weaned calf crop[b]	
Percent	Kilos	Percent	Kilos
1	221	1	221
2	442	2	442
3	663	3	663
4	867	4	867
5	1,114		
6	1,335		
7	1,547		
8	1,777		

Improved cow milking performance[b]		Increased calving percent[c]	
Percent	Kilos	Percent	Kilos
1	221	1	210
2	442	2	420
3	663	3	630
4	867	4	824
		5	1,058
		8	1,648
		10	2,116

Greater proportion on males calves[d]	
Percent	Kilos
1	221
2	442
3	663
4	867

[a]Beef breed weaned calf weight is 261 kg while the average weight of dairy calves entering the feedlot is 264 kg. Thus, the conversions in this table are used for both systems.

[b]Based on a weaned calf weight of 261 kg (see Chapter 5) and a weaned 85 percent calf crop. Thus, for example, 261 kg times 0.02 = 5.2 kg per calf times 85 calves per 100 cows = 442 kg.

[c]Same as increased weaned calf crop, but adjusted for 5 percent death loss from birth to weaning. Thus, for example, at the 2 percent level, 442 kg x 0.95 = 420 kg.

[d]Assume a 20 percent difference between male and female calves. With an average of 261 kg per calf, then females would weigh 235 kg and males 287 kg. Using weighted averages:

$$(2.35 \times 0.50) + (287 \times 0.50) =$$

$$117 \quad + \quad 144 \quad = 261$$

The weighted average for a 10 percent increase in male calves would be:

$$(235 \times 0.40) + (2.87 \times 0.60) =$$

$$94 \quad + \quad 172 \quad = 266$$

Then, 266-261 = 5 kg per calf x 85 calves = 425 kg or about the same as the 2 percent level of increased weaned calf weight.

Appendix 8.6. Calculations of beef production per head of cattle inventory in Japan, 1990, 2000 and 2010

Item	Cows		Feedlots		Other	
	Beef breed	Dairy breed	Beef breed	Dairy breed	inventory[d]	Total
Inventory, 1984 (Head)[a]	678,000	1,101,000	571,500	913,900	1,417,600	4,682,000
Mean productivity increases (Percent)[b]						
1990	10	05	05	02	—	—
2000	28	22	13	06	—	—
2010	43	30	18	08	—	—
Low productivity increases (Percent)[b]						
1990	04	01	02	00	—	—
2000	17	13	08	03	—	—
2010	30	20	12	05	—	—
High productivity increases (Percent)[b]						
1990	15	09	08	04	—	—
2000	39	31	18	09	—	—
2010	55	39	24	11	—	—

Continued

Appendix 8.6. Calculations of beef production per head of cattle inventory in Japan, 1990, 2000 and 2010 (continued)

Item	Cows		Feedlots		Other	
	Beef breed	Dairy breed	Beef breed	Dairy breed	inventory[d]	Total
Potential inventory reduction to achieve same total production obtained in 1984 (Head)[c]						
Mean						
1990	610,200	1,045,950	542,925	895,622	1,417,600	4,512,297
2000	488,160	858,780	440,055	859,066	1,417,600	4,063,661
2010	386,460	770,700	468,630	840,788	1,417,600	3,884,178
Low						
1990	650,880	1,089,990	560,070	913,900	1,417,600	4,632,440
2000	562,740	847,770	525,780	886,483	1,417,600	4,240,373
2010	474,600	880,800	502,920	868,205	1,417,600	4,144,125
High						
1990	576,300	1,001,910	525,780	877,344	1,417,600	4,398,934
2000	413,580	759,690	468,630	831,649	1,417,600	3,891,149
2010	305,100	671,610	434,340	813,371	1,417,600	3,642,021
Total beef production 1984 (1,000 MT)						510
Production per head of inventory (kg)						
Mean						
1990						122.8
2000						125.5
2010						131.3
Low						
1990						110.1
2000						120.3
2010						123.1
High						
1990						115.9
2000						131.1
2010						140.0

[a]February 1, 1984. See Tables 3.4 and 3.5. Beef breed cows are the category "female, 2 years and older." Dairy breed cows are the category "number of cows that have calved and are in lactation." Beef breed in feedlots are the sum of the category "males." Dairy breeds in feedlots are the category "dairy cattle for beef."

[b]Feedlots are for growth rate only.

[c]Inventory in 1984 multiplied by difference between percentage productivity increase and 100 percent.

[d]From Table 3.1.

[e]From Table 8.1.

Appendix 8.7. Calculations for Tables 8.4, beef breed feedlot cost per slaughter animal in constant 1983 terms, Japan, 1990, 2000 and 2010

	Cash expenses (variable)			Fixed expenses	Total cash
	Feed	Nonfeed	Calf		
	- - - - - - - - - - - - - Yen - - - - - - - - - - - - -				
1983 costs[a]	241,516	21,835	346,390	24,362	606,195
	- - - - - - - - - - - - Percent - - - - - - - - - - - -				
Cost reduction[b]					
Mean scenario					
1990	13	06	10	06	—
2000	31	12	28	12	—
2010	42	16	43	16	—
Low scenario					
1990	06	03	04	03	—
2000	19	08	17	08	—
2010	28	11	30	11	—
High scenario					
1990	20	09	15	09	—
2000	52	16	39	16	—
2010	55	20	55	20	—
	- - - - - - - - - - - - - Yen - - - - - - - - - - - - -				
Adjusted costs					
Mean scenario					
1990	210,119	20,525	311,751	21,672	536,159
2000	166,646	19,215	249,401	20,288	427,642
2010	140,079	18,341	197,442	19,366	347,320
Low scenario					
1990	227,025	21,180	332,534	22,363	575,194
2000	195,628	20,088	287,504	21,211	524,431
2010	173,892	19,433	242,473	20,519	428,409
High scenario					
1990	193,213	19,870	294,432	20,980	500,587
2000	115,928	18,341	211,298	19,366	337,025
2010	108,682	17,468	155,876	18,444	272,562

[a]From Table 6.2.

[b]Percentages from Appendix 8.3. Feed includes all three categories "growth rate," "feed efficiency," and "other economic efficiency." Nonfeed is the termed "other economic efficiency;" fixed expenses are from the category "other economic efficiency." Calf reduction from Appendix 8.1. Totals adjusted for manure credit of ¥27,908.

Appendix 8.8. Calculations for Tables 8.5, dairy breed feedlot cost per slaughter animal in constant 1983 terms, Japan, 1990, 2000 and 2010

	Basic production costs			Minus by-product value	Net basic production cost
	Feed	Nonfeed	Calf		
	Yen per steer sold				
1983 costs[a]	205,950	19,123	180,674	8,584	397,163
	Percent				
Cost reduction[b]					
Mean scenario					
1990	11	08	05	—	—
2000	23	15	22	—	—
2010	28	18	30	—	—
Low scenario					
1990	05	04	01	—	—
2000	15	10	13	—	—
2010	20	13	20	—	—
High scenario					
1990	17	11	09	—	—
2000	31	19	31	—	—
2010	36	22	39	—	—
	Yen				
Adjusted costs					
Mean scenario					
1990	189,296	17,976	171,640	8,584	370,328
2000	158,582	16,255	140,926	8,584	307,179
2010	148,284	15,681	126,472	8,584	281,853
Low scenario					
1990	195,653	18,358	178,867	8,584	384,294
2000	175,058	17,211	157,186	8,584	340,871
2010	164,760	16,637	144,539	8,584	317,352
High scenario					
1990	170,939	17,019	164,413	8,584	343,787
2000	142,106	15,490	124,665	8,854	273,677
2010	131,808	14,916	110,211	8,584	248,351

[a]From Appendix 6.9. Totals for 50 head converted to cost per animal sold by dividing by 49.5 head (to account for death loss).

[b]Percentages from Appendix 8.3. Feed includes all three categories "growth rate," "feed efficiency," and "other economic efficiency;" nonfeed is the term "other economic efficiency." Calf reduction from Appendix 8.2

Appendix 9.1. Beef and veal production, trade and consumption, calendar year basis, Japan, 1975-1985

Calendar year	Production	Imports	Consumption Total[a]	Consumption Per capita
	- - - - - -	1,000 MT	- - - - - -	- kg -
1975	353	64	411	3.7
1976	298	130	436	3.9
1977	360	120	465	4.2
1978	403	135	427	4.7
1979	402	185	541	5.0
1980	418	174	590	5.1
1981	471	174	625	5.3
1982	481	174	655	5.5
1983	495	196	691	5.8
1984[b]	535	210	730	6.1
1985[c]	505	225	760	6.3

Source: U.S. Department of Agriculture, Foreign Agriculture Circular, Livestock and Meat, and Livestock and Poultry Situation, various issues.

[a]Adjusted for exports and changes in stocks.

[b]Preliminary.

[c]Forecast.

Appendix 10.1. Japanese beef imports by countries 1965-1983

Year	Sweden	Canada	U.S.A.	Mexico	Australia	New Zealand	Total[a]
	- - - - - - -	- - - - - -	- - - - -	- - Metric	tons - -	- - - - - -	- - - - - -
1965	--	--	7	--	7,774	2,569	10,813
1970	--	25	362	--	20,123	2,511	23,227
1975	--	1	3,545	18	3,709	3,512	44,923
1976	--	367	11,864	106	77,025	4,639	94,233
1977	--	284	7,330	803	72,055	3,903	84,547
1978	--	333	13,026	1,342	78,173	7,800	100,864
1979	--	819	24,672	1,447	101,268	3,510	131,792
1980	--	1,579	23,674	903	93,614	3,991	123,952
1981	60	1,959	27,543	773	97,071	6,148	123,646
1982	249	69	32,079	442	86,099	3,645	122,694
1983	133	232	37,728	17	91,043	7,734	134,542

Source: MAFF. The Meat Statistics in Japan. Tokyo, January 1, 1985.

[a]Includes the other minor suppliers.

Appendix 10.2. Calculated proportion of Japanese beef imports by countries 1965-1983

Year	Sweden	Canada	U.S.A.	Mexico	Australia	New Zealand	Total[a]
				Percent			
1965	--	--	0.0	--	71.9	23.8	100
1970	--	0.1	1.6	--	86.6	10.8	100
1975	--	0.0	7.9	0.0	82.6	7.8	100
1976	--	0.4	12.6	0.1	81.7	4.9	100
1977	--	0.3	8.7	0.9	85.2	4.6	100
1978	--	0.3	12.9	1.3	77.5	7.7	100
1979	--	0.6	18.7	1.1	76.8	2.7	100
1980	--	1.3	19.1	0.7	75.4	3.2	100
1981	0.0	1.6	22.3	0.6	70.4	5.0	100
1982	0.2	0.1	26.1	0.4	70.2	3.0	100
1983	0.1	0.2	27.4	0.0	66.2	5.6	100

Source: Derived from Appendix 10.1.

Index

9780367289584